CLIFFS

GED Science Test

PREPARATION GUIDE

The New High School
Equivalency Examination

by
Harold D. Nathan, Ph.D.

Series Editor
Jerry Bobrow, M.A.

Consultants
Debbie Moran, M.A.
Merritt L. Weisinger, J.D.

LINCOLN, NEBRASKA   68501

ACKNOWLEDGMENTS

Credit is due the following authors for information and passages from several Cliffs Rapid Reviews.

Stewart Brooks: *Biology* (1968)
William Caldwell: *Chemistry* (1968)
Richard Pearl: *Geology* (1968)
Joan Rahn: *Botany* (1969) and *Microbiology* (1974)

Illustrations by Kirk Greuter.

Thanks are also due Michele Spence of Cliffs Notes for her alert, patient editing.

ISBN 0-8220-2010-6
© Copyright 1980

by

Harold D. Nathan

All Rights Reserved
Printed in U.S.A.

CONTENTS

Preface vii
Study Guide Checklist viii

PART I: INTRODUCTION

FORMAT OF THE NEW GED SCIENCE TEST 3
GENERAL DESCRIPTION 3
QUESTIONS COMMONLY ASKED ABOUT THE NEW GED TEST BATTERY 4
A SYSTEMATIC OVERALL APPROACH 6

PART II: ANALYSIS OF EXAM AREAS

HOW TO TAKE THE TEST 9
    Ability Tested • Basic Skills Necessary • Directions • Analysis of Directions • Suggested Approach with Samples • Additional Tips

PART III: ASSESSMENT

MINI-TEST WITH COMPLETE ANSWERS AND EXPLANATIONS 15
    Answer Sheet for the Mini-Test 17
    Mini-Test 19
        Part A: General Science Knowledge 19
        Part B: Science Reading Comprehension 20
    Answer Key for the Mini-Test 27
    Scoring Your GED Science Test 28
    Mini-Test: Score Approximator 28
    Analyzing Your Test Results 28
    Mini-Test: General Analysis Sheet 29
    Mini-Test: Subject Area Analysis Sheet 29
    Analysis—Tally Sheet for Questions Missed 30
    Complete Answers and Explanations for the Mini-Test 31

Part A: General Science Knowledge 31
Part B: Science Reading Comprehension 32

PART IV: SUBJECT AREA REVIEWS

HOW TO REVIEW 37
BIOLOGY REVIEW 38
  Basic Concepts 38
    Biology 38
    The Cell 38
    Metabolism 39
    Mitosis 39
    Plants 40
    Photosynthesis 41
    Animals 42
    The Sensory System 44
    The Nervous System 44
    The Digestive System 45
    The Circulatory System 46
    Reproduction 47
    Nucleic Acids 47
    Genes 48
    Evolution 49
    Taxonomy 49
    Ecology 50
  Glossary of Terms in Biology 50
GEOLOGY REVIEW 57
  Basic Concepts 57
    Geology 57
    The Earth's Structure 57
    The Rock Cycle 58
    Minerals 59
    Igneous Rocks 60
    Sedimentary Rocks 61
    Metamorphic Rocks 61
    Weathering 62
    Erosion 62
    Strata 63
    Fossils 63
    The Geological Time Scale 63
    Oceans 64

Continents	64
Earth Movements	65
Natural Resources	65
Meteorology	66
Glossary of Terms in Geology and Meteorology	67
CHEMISTRY REVIEW	**72**
Basic Concepts	72
Chemistry	72
Atoms	72
Subatomic Particles	73
Chemical Elements	74
Abridged Periodic Table of the Chemical Elements	75
Compounds	76
Bonds	76
Solids	76
Liquids	77
Gases	77
Water	78
Solution	79
Ions	79
Acids and Bases	79
Electrolysis	80
Chemical Reactions	81
Glossary of Terms in Chemistry	81
PHYSICS REVIEW	**86**
Basic Concepts	86
Physics	86
Measurement	86
Motion	87
Newton's Laws	87
Gravitation	88
Energy	88
Temperature	89
Sound	90
Electricity	90
Magnetism	91
Light	91
Relativity	93
Nuclear Energy	94
Astronomy	95
Glossary of Terms in Physics and Astronomy	96

PART V: PRACTICE-REVIEW-ANALYZE-PRACTICE
Two Full-Length Practice Tests

PRACTICE TEST 1 WITH COMPLETE ANSWERS AND EXPLANATIONS 103
   Answer Sheet for Practice Test 1 105
   Practice Test 1 107
      Part A: General Science Knowledge 107
      Part B: Science Reading Comprehension 110
   Answer Key for Practice Test 1 121
   Scoring Your GED Science Test 122
   Practice Test 1: Score Approximator 122
   Analyzing Your Test Results.. 122
   Practice Test 1: General Analysis Sheet 123
   Practice Test 1: Subject Area Analysis Sheet 123
   Analysis—Tally Sheet for Questions Missed 124
   Complete Answers and Explanations for Practice Test 1 125
      Part A: General Science Knowledge 125
      Part B: Science Reading Comprehension 127

PRACTICE TEST 2 WITH COMPLETE ANSWERS AND EXPLANATIONS 131
   Answer Sheet for Practice Test 2 133
   Practice Test 2 135
      Part A: General Science Knowledge 135
      Part B: Science Reading Comprehension 138
   Answer Key for Practice Test 2 150
   Scoring Your GED Science Test 151
   Practice Test 2: Score Approximator 151
   Analyzing Your Test Results.. 151
   Practice Test 2: General Analysis Sheet 152
   Practice Test 2: Subject Area Analysis Sheet 152
   Analysis—Tally Sheet for Questions Missed 153
   Complete Answers and Explanations for Practice Test 2 154
      Part A: General Science Knowledge 154
      Part B: Science Reading Comprehension 157

Final Preparation: "The Final Touches"... 161
Appendix: Policies of State Departments of Education and Official
   GED Centers 162

PREFACE

We know that passing the GED Test Battery is important to you! And thorough preparation is the key to doing your best. Because of this, your study time must be used most effectively. With this purpose in mind, the Cliffs GED Test Preparation Series was developed by leading experts in the field of test preparation. These guides are the most comprehensive test preparation materials that you can realistically complete in a reasonable time. Each of the GED Guides is easy to use, direct, precise, and thorough, giving you the important information, tips, and strategies that you need to pass the GED. These strategies and techniques have been carefully tested and are presently used in test preparation programs at many leading universities and colleges.

This GED Guide is divided into five parts:

PART I: Introduction—a general description of the exam, recent format, questions commonly asked, and basic overall strategy.
PART II: Analysis of Exam Areas—focuses on ability tested, basic skills necessary, directions, analysis of directions, suggested approaches with samples, and additional tips.
PART III: Assessment—a half-length Mini-Test to assess your strengths and weaknesses.
PART IV: Subject Area Reviews—intensive reviews of the important areas (biology, geology, chemistry, and physics), including an introduction on how to review and a 500-word glossary.
PART V: Practice-Review-Analyze-Practice—two complete full-length practice tests with answers and *in-depth explanations*.

The Mini-Test and each practice test are followed by analysis charts and score approximators to assist you in evaluating your progress.

These guides are not meant to substitute for comprehensive courses, but if you follow the Study Guide Checklist and study regularly, you will get the best test preparation possible.

STUDY GUIDE CHECKLIST

____ 1. Read the GED information materials available at your local GED center. Check minimum score, age, and residence requirements.
____ 2. Become familiar with the Test Format, page 3.
____ 3. Read the General Description and Questions Commonly Asked about the GED Test Battery, starting on page 3.
____ 4. Learn the techniques of A Systematic Overall Approach, page 6.
____ 5. Carefully read Part II, Analysis of Exam Areas, starting on page 9.
____ 6. Take the Mini-Test Assessment, starting on page 15.
____ 7. Check your answers and analyze your results, page 27.
____ 8. Fill out the Tally Sheet for Questions Missed to pinpoint your mistakes, page 30.
____ 9. Read How to Review, page 37.
____ 10. Review Biology (basic concepts), starting on page 38.
____ 11. Go over Glossary of Terms in Biology, starting on page 50.
____ 12. Review Geology (basic concepts), starting on page 57.
____ 13. Go over Glossary of Terms in Geology and Meteorology, starting on page 67.
____ 14. Review Chemistry (basic concepts), starting on page 72.
____ 15. Go over Glossary of Terms in Chemistry, starting on page 81.
____ 16. Review Physics (basic concepts), starting on page 86.
____ 17. Go over Glossary of Terms in Physics and Astronomy, starting on page 96.
____ 18. Strictly observing time allotments, take Practice Test 1, beginning on page 103.
____ 19. Check your answers and analyze your Practice Test 1 results, page 121.
____ 20. Fill out the Tally Sheet for Questions Missed to pinpoint your mistakes, page 124.
____ 21. While referring to each item of Practice Test 1, study ALL the Answers and Explanations that begin on page 125.
____ 22. Review weak areas as necessary.
____ 23. Strictly observing time allotments, take Practice Test 2, beginning on page 131.

STUDY GUIDE CHECKLIST   ix

___ 24. Check your answers and analyze your Practice Test 2 results, page 150.
___ 25. Fill out the Tally Sheet for Questions Missed to pinpoint your mistakes, page 153.
___ 26. While referring to each item of Practice Test 2, study ALL the Answers and Explanations that begin on page 154.
___ 27. Review weak areas.
___ 28. Review How to Take the Test, starting on page 9.
___ 29. Carefully read Final Preparation: "The Final Touches," page 161.

＃ PART I: Introduction

FORMAT OF THE NEW GED SCIENCE TEST

Time: 90 Minutes   Total Questions: 60

Part A: Background Knowledge	20 Questions
Part B: Reading Comprehension	40 Questions

CONTENT AREAS

1. Biology	50%
2. Geology	20%
3. Chemistry	15%
4. Physics	15%

GENERAL DESCRIPTION

The GED Science Test lasts 1½ hours (90 minutes) and consists entirely of multiple-choice questions. The test contains 60 questions that are of two basic types: general science knowledge (20 questions) and science reading comprehension (40 questions). The areas covered are biology, geology, chemistry, and physics, with biology comprising 50% of the exam and the physical sciences comprising the other 50%. The test is scored from 20 to 80 with all questions being of equal value.

QUESTIONS COMMONLY ASKED ABOUT THE GED TEST BATTERY

Q: Who administers the GED Test Battery?
A: The GED Test Battery is administered by the GED Testing Service of the American Council on Education (ACE). The test development was conducted by Educational Testing Service with the guidelines drawn up by experienced secondary school educators.

Q: When are the GEDs given?
A: The tests are administered nationwide on a continuous basis. You can get administration dates and test locations by contacting your local GED testing centers, high schools, or adult schools.

Q: Do I have to take the complete battery of tests at once?
A: No! You may take one, two, three, four, or all five of the tests on the same day.

Q: Can I take the GEDs more than once?
A: Yes! You may take any or all of the tests more than once. But remember, your plan is to pass on the *first* try.

Q: Are there any special editions of the GED?
A: Yes! In addition to the English version, the GED is also given in Spanish and French. There are also special large-print and/or tape-recorded editions for handicapped candidates.

Q: What is a passing score?
A: There are no national standards for the GEDs. Each state has established its own standards for passing the tests. In many cases you need only 40% correct to pass. The Appendix includes a chart to give you an indication of previous requirements. Each candidate should check the up-to-date policies in his or her specific state. It should be noted that some states require candidates to pass each test, while others require the average score of the tests to be passing.

Q: What grade level are the tests?
A: The tests are standardized to reflect the range of knowledge and ability of twelfth grade students who are certain to graduate.

Q: How are the *new* GEDs different from the *old* GEDs?
A: The new GED Test Battery was shortened from 10 hours to 6 hours with fewer questions in each test. The Social Studies and Science Tests now include general knowledge questions and shorter reading passages. The

Writing Skills Test, formerly the Test on Correctness and Effectiveness of Expression, now includes logic (not formal logic) and organization questions. The Reading Skills Test now draws more information from modern literature and everyday reading. The Math Test is basically the same.

Q: How should I prepare?
A: Understanding and practicing test-taking strategies will help a great deal, especially on the reading comprehension type questions. Subject matter reviews, by reviewing major concepts and important terminology, are invaluable. Both subject matter and strategies are covered in this book. Also, many states offer preparation programs to assist students. Check with your local high school, adult school, or occupational center for further information.

Q: Should I guess on the tests?
A: Yes! Since there is no penalty for guessing, GUESS if you have to. If possible, first try to eliminate some of the choices to increase your chances of choosing the right answer.

Q: How and when should I register and where can I get more information?
A: This information is available from your local GED Testing Center, or write to General Educational Development, GED Testing Service of the American Council on Education, One Dupont Circle, Washington, D.C., 20036.

A SYSTEMATIC OVERALL APPROACH

Many who take the GED don't get the scores that they deserve because they waste time on hard questions, leaving insufficient time to answer the easy questions. Don't let that happen to you. Use the following system to mark your answer sheet:

1. Answer easy questions immediately.
2. Mark a "+" next to the number of any question that seems solvable but very difficult. Go on to the next question.
3. Mark a "−" next to any question that seems impossible. Go on to the next question.

The strategy is to defer difficult and impossible questions. Act quickly. Don't waste time deciding whether a question is a "+" or a "−."

Your answer sheet should look something like this after you finish answering your easy questions:

   1. ① ② ● ④ ⑤
   2. ● ② ③ ④ ⑤
+3. ① ② ③ ④ ⑤
   4. ① ② ③ ④ ●
−5. ① ② ③ ④ ⑤

Then go back and answer your difficult (+) questions. By now they may seem easier, after your mind has worked on other questions. Finally, use educated guesses to answer the impossible (−) questions. Remember, there is no penalty for guessing and it is unwise to leave any question unanswered.

Make sure to erase your "+" and "−" marks just before the end of the test, as stray marks on the answer sheet can confuse the scoring machine.

Practice this Systematic Overall Approach on the tests in this book, so it becomes a habit for the actual GED Science Test.

PART II: Analysis of Exam Areas

This section is designed to introduce you to the GED Science Test by carefully reviewing the

1. Ability Tested
2. Basic Skills Necessary
3. Directions
4. Analysis of Directions
5. Suggested Approach with Samples

The emphasis in this section is on test-taking strategies and techniques.

HOW TO TAKE THE TEST

The GED Science Test contains about 60 questions, based either on reading passages or on background knowledge of scientific concepts.

Ability Tested

The GED Science Test tests your general knowledge of high school science and your ability to read, comprehend, and interpret science reading passages. Your ability to apply knowledge is also tested.

Basic Skills Necessary

The test will include passages and questions from biology, geology, chemistry, and physics. Some simple questions on meteorology and astronomy may also appear. Biology is heavily emphasized, making up approximately 50 percent of all questions. You are not expected to be familiar with all topics on the test, but you must demonstrate that you understand the basics of high school science.

More specifically, the biology questions concern cells, plants, and animals. You are expected to understand the main processes of life: photosynthesis, metabolism, reproduction, and evolution. You should know the names and descriptions of major plant and animal groups. Human anatomy is stressed on the test, covering the senses, the nervous system, the circulation of blood, the respiratory system, and the digestive system. Other topics in biology include genetics, diseases, and ecology.

Geology subjects you are responsible for are earth structure, rock types, volcanoes, sedimentation, and glaciers. You may also be questioned on earth history, including fossils, mountain building, and continental drift. The atmosphere and weather are also considered.

In chemistry, the GED covers the familiar chemical elements and the nature of chemical reactions. You may be asked about substances like subatomic particles, atoms, molecules, solutions, and mixtures. You are expected to understand the behavior of gases and the principle of the conservation of matter.

Physics questions require you to understand measurement systems, both English and metric. You may be asked about the laws of motion, the conservation of energy, heat transfer, optical effects, electricity, and magnetism. The GED may also test you on elementary astronomy, particularly the solar system.

Directions

The GED Science Test consists of 60 multiple-choice questions in two parts. In the first part are 20 questions based on background knowledge of scientific concepts. The second part contains 40 questions based on several brief reading passages. You have 90 minutes to complete both parts of this test. There is no penalty for guessing.

Part A: General Science Knowledge

These questions are not based on a reading passage. Select the best answer based on your knowledge of the concepts of natural science.

Part B: Science Reading Comprehension

This section contains 10 reading passages, with several questions about each passage. Read the passage first and then answer the questions following it. Refer to the passage as often as necessary in selecting the best answer. Remember, there is no penalty for guessing.

Analysis of Directions

Although the first part (A) of the test contains the short questions based on background knowledge, it is best to skip immediately to the second part (B) with reading passages. The reading questions are easier and solving them will allow your confidence to build up. Also, some of the reading passages may jog your memory and enable you to correctly answer conceptual questions (Part A) which you would have missed. So plan on working on the reading passages (Part B) for the first 60 minutes, then returning to the conceptual questions (Part A) for the final 30 minutes. There is ample time to complete everything, as long as you work steadily.

Try to follow that sequence in the practice tests in this book. Impose a time limit on yourself and take the test as if it were the real GED. Teach yourself to concentrate on each question and seek the best answer. The good habits you acquire during practice will help you during the actual test.

Notice that there is no penalty for guessing. Therefore you should not omit an answer for any question; even blind guesses will be correct about 20 percent of the time (5 multiple choices) and should increase your score. But don't guess blindly when you first encounter a difficult question. The correct answer may pop into your mind later, perhaps while reading passages or working on related questions. It is essential to defer blind guessing until near the end of the test. Use the Systematic Overall Approach described at the end of Part I.

Suggested Approach with Samples

First, skip to Part B with the reading passages. Glance at the passage to determine which branch of science it concerns, then quickly skim through the questions (not answers) to find what you will be asked. As you carefully read the passage, look for references to the questions. The long, technical words in a question usually are to be found in the passage near the answer to that question. Often there are clues to the correct answer in the question as well as in the passage.

In Part B there are two types of questions, reading questions and thinking questions, which are about equally frequent. The reading questions are easier, for their answers are *printed* in the passages. Thinking questions require you to *interpret* the passage, drawing a conclusion based on material in the passage and some background knowledge. Let's look at a sample reading passage and questions.

Simple fruits may be fleshy or dry. Simple fleshy fruits are those which have moist, often juicy, pericarps; simple dry fruits have woody or papery pericarps. Simple fleshy fruits are either drupes, berries, or pomes. A drupe (peach, olive) comes from a simple ovary with one ovule; the exocarp and mesocarp are fleshy, but the endocarp is stony, forming a pit. A berry (tomato, grape) differs from a drupe in being derived from a compound ovary, in containing numerous seeds, and in having a completely fleshy pericarp. Some special types of berries are the hesperidium (orange, lemon) which has a leathery rind and the pepo (squash, watermelon) which has a hard rind. A pome (apple, pear) has an edible portion composed largely of floral parts that surround the ovary. The core is formed from the endocarp.

1. Which fruit comes from a simple ovary?
   (1) apple
   (2) lemon
   (3) tomato
   (4) olive
   (5) watermelon

2. Which fruit is *not* a berry?
   (1) grape
   (2) orange
   (3) pear
   (4) squash
   (5) tomato

3. Which fruit is a hesperidium?
   (1) apricot
   (2) avocado
   (3) plum
   (4) pear
   (5) grapefruit

Question 1 is a very simple reading question. The key phrase in the question is *simple ovary* and the pertinent sentence in the passage states that "a drupe (peach, olive) comes from a simple ovary with one ovule." Therefore the correct answer is (4) olive.

Question 2 is a slightly more difficult reading question. The key word is *berry,* and reference to that word in the passage informs you that typical berries are the tomato and grape, while special berries are the orange, lemon, squash, and watermelon. The correct answer is, then, (3) pear.

Question 3 is a thinking question, requiring you to think beyond the passage. The key technical word in the question is *hesperidium* and the passage states that examples are the orange and lemon, neither of which is one of the 5 choices for an answer. However, you are aware that oranges and lemons are citrus fruits, so another citrus fruit would be a hesperidium. Of the possible answers, only (5) grapefruit is a citrus fruit, so that choice is correct.

After you have finished the reading passages, return to Part A and answer the conceptual questions from your background knowledge. Here's a sample question.

4. Which substance is *not* gaseous at normal temperature and pressure?
    (1) ammonia
    (2) chlorine
    (3) methane
    (4) nitrogen
    (5) sulfur

You may know that sulfur is a yellow solid, so (5) is the correct choice. Let's assume, however, that you're not sure about sulfur. You can attack the question by eliminating wrong choices. Ammonia is a pungent gas used in cleaning solutions. Chlorine is a green, poisonous gas which was employed in World War I. Methane is marsh gas, a combustible natural gas from decay. Nitrogen is the principal component of air. Knowing any of these four facts enables you to eliminate some choices and make an educated guess.

Additional Tips

Take your practice tests seriously and develop good test-taking habits. Don't rush through a test to "get it over with," but use all the time allowed.

For each question you miss, read the explanation carefully. In that way you will learn more science and avoid that type of mistake in the future.

Don't get discouraged. Learning can be difficult. If you study and practice earnestly, your performance will improve. And on the actual GED test don't be alarmed at strange questions. Nobody expects you to answer all of them correctly. Your strategy is to keep working and get the ones you know how to answer.

PART III: Assessment

The Mini-Test that follows is designed to assess your strengths and weaknesses. This assessment includes complete answers and explanations. You have 45 minutes to complete the Mini-Test.

MINI-TEST
WITH COMPLETE ANSWERS AND EXPLANATIONS

ANSWER SHEET FOR THE MINI-TEST
(Remove This Sheet and Use It to Mark Your Answers)

1 ① ② ③ ④ ⑤	16 ① ② ③ ④ ⑤
2 ① ② ③ ④ ⑤	17 ① ② ③ ④ ⑤
3 ① ② ③ ④ ⑤	18 ① ② ③ ④ ⑤
4 ① ② ③ ④ ⑤	19 ① ② ③ ④ ⑤
5 ① ② ③ ④ ⑤	20 ① ② ③ ④ ⑤
6 ① ② ③ ④ ⑤	21 ① ② ③ ④ ⑤
7 ① ② ③ ④ ⑤	22 ① ② ③ ④ ⑤
8 ① ② ③ ④ ⑤	23 ① ② ③ ④ ⑤
9 ① ② ③ ④ ⑤	24 ① ② ③ ④ ⑤
10 ① ② ③ ④ ⑤	25 ① ② ③ ④ ⑤
11 ① ② ③ ④ ⑤	26 ① ② ③ ④ ⑤
12 ① ② ③ ④ ⑤	27 ① ② ③ ④ ⑤
13 ① ② ③ ④ ⑤	28 ① ② ③ ④ ⑤
14 ① ② ③ ④ ⑤	29 ① ② ③ ④ ⑤
15 ① ② ③ ④ ⑤	30 ① ② ③ ④ ⑤

CUT HERE

MINI-TEST

Time: 45 Minutes
30 Questions

DIRECTIONS

This introductory test consists of 30 multiple-choice questions in two parts. In first part are 10 questions based on background knowledge of scientific concepts. The second part contains 20 questions based on several brief reading passages. You have 45 minutes to complete both parts of this test. There is no penalty for guessing.

Part A: General Science Knowledge

The following questions are not based on a reading passage. Select the best answer based on your knowledge of the concepts of natural science.

1. Which of the following organs is *not* part of the circulatory system?
   (1) aorta
   (2) capillary
   (3) heart
   (4) trachea
   (5) vein

2. Litmus paper dipped into an aqueous solution would detect the presence of
   (1) alcohol
   (2) ammonia
   (3) colloids
   (4) oxygen
   (5) sugar

3. A light-year is a unit of
   (1) distance
   (2) energy
   (3) illumination
   (4) temperature
   (5) time

4. All of the following processes directly affect the concentration of carbon dioxide in the atmosphere *except*
   (1) combustion of oil
   (2) decay of forest waste
   (3) photosynthesis of plants
   (4) respiration of animals
   (5) rusting of iron

5. An instrument used to record earthquakes is called a
   (1) barometer
   (2) bathyscaphe
   (3) microscope
   (4) seismograph
   (5) spectrometer

6. An eagle, elephant, frog, and shark are all
   (1) amphibians
   (2) carnivores
   (3) mammals
   (4) terrestrial
   (5) vertebrates

7. If the reaction of hydrogen and oxygen to yield water vapor

   $$2H_2 + O_2 \rightarrow 2H_2O$$

   yields 12 liters of water vapor, how many liters of gases reacted?
   (1) 6   (2) 8   (3) 12   (4) 18   (5) 24

8. Were we to perform an experiment measuring the speeds of pieces of aluminum and copper falling in a vacuum, we should observe that
   (1) the aluminum falls faster
   (2) the copper falls faster
   (3) the larger piece falls faster
   (4) the rounder piece falls faster
   (5) the pieces fall at equal speeds

9. Crossing two orange marigold plants, each an orange/yellow hybrid with orange dominant and yellow recessive, should yield what percentage of orange-appearing offspring?
   (1) none, unless a mutation has taken place
   (2) 25%
   (3) 50%
   (4) 75%
   (5) 100%

10. All of the following features are likely to be associated with a large river *except*
    (1) delta
    (2) moraine
    (3) floodplain
    (4) meanders
    (5) tributaries

Part B: Science Reading Comprehension

This section contains 5 reading passages, with several questions about each passage. Read the passage first and then answer the questions following it. Refer to the passage as often as necessary in selecting the best answer. Remember, there is no penalty for guessing.

There is believed to be more hydrogen in the universe than any other chemical element. A high percentage occurs in the sun's atmosphere. It is found in small amounts in the earth's atmosphere; because of its low density, the light hydrogen molecules escape the earth's relatively weak gravitational field. In combined form, hydrogen occurs in water, in acids and bases, and in most organic compounds. Plant and animal tissues contain hydrogen in combination with carbon, oxygen, nitrogen, and sulfur.

The purest hydrogen is prepared from the electrolysis of water. However, this process is expensive because of the electrical energy required. Hydrogen is prepared commercially by passing steam over hot carbon:

$$H_2O + C \rightarrow CO + H_2$$

If the hydrogen is desired alone, it must be separated from the carbon monoxide formed in the reaction. A mixture of hydrogen and carbon monoxide is known as water gas, which is used as an industrial fuel. A considerable amount of hydrogen is the by-product of the catalytic cracking of petroleum hydrocarbons. When heated to a high temperature with steam, and with the aid of a catalyst, natural gas (methane) yields hydrogen:

$$\underset{\text{methane steam}}{CH_4 + H_2O} \xrightarrow{\text{catalyst}} CO + 3H_2$$

11. Hydrogen escapes from the air into outer space because
    (1) it combines with carbon and other elements
    (2) its molecules have very low mass
    (3) the earth has a gravitational field
    (4) there is more hydrogen in the universe than any other chemical element
    (5) water is very abundant

12. In the method of preparing the purest hydrogen what other substance might be produced at the same time?
    (1) carbon   (4) methane
    (2) carbon dioxide   (5) oxygen
    (3) carbon monoxide

13. The commercial process yields
    (1) a mixture of two gases   (4) one solid and one gas
    (2) one liquid and one gas   (5) two liquids
    (3) one pure gas

14. In the second equation the source of the final hydrogen is the
    (1) carbon monoxide
    (2) catalyst
    (3) methane
    (4) steam
    (5) methane and steam

Wine is the product of alcoholic fermentation of the juice of fruits or other vegetable materials. Most commonly it is produced from ripe grapes rich in fermentable sugars. Grape juice is inoculated with *Saccharomyces ellipsoideus* (for commercial wines) or with wild yeast from grape leaves and the surface of the fruit (for homemade wines). Fermentation is allowed to proceed aerobically for several days while the yeasts reproduce rapidly. Fermentation then proceeds anaerobically for 1 to 2 weeks at 25 to 30°C. The wine is then stored in wooden casks; as it ages, aromas and flavors develop. Wine is bottled, pasteurized, and usually stored for several years before marketing. Pasteurization prevents growth of organisms, especially *Lactobacillus* and *Acetobacter*, which produce lactic acid and acetic acid, respectively. The alcohol content of wine varies from 7 to 15 percent. A dry wine is one in which the sugar has been fermented completely; in a sweet wine, some of the sugar remains. Brandy is a distillate of wine.

15. Fermentation occurs
    (1) first with air, then without it
    (2) first without air, then with it
    (3) only with air
    (4) only without air
    (5) with air only if the wine is commercial

16. Which of the following substances and qualities decreases as fermentation progresses?
    (1) alcohol
    (2) flavor
    (3) sugar
    (4) temperature
    (5) yeast

17. Which of the following substances is a yeast?
    (1) *Acetobacter*
    (2) brandy
    (3) grape leaves
    (4) *Lactobacillus*
    (5) *Saccharomyces*

18. What is the source of yeast for home fermentation?
    (1) grapes
    (2) water
    (3) pasteurization
    (4) sugar
    (5) wooden casks

Carbonic acid is produced by the addition of carbon dioxide from air and vegetation to water. The presence of carbonic acid makes groundwater a good solvent of carbonate rock, especially limestone. Sinkholes are produced at the surface, caverns are formed underground, and a karst topography (full of holes) results if the process continues long enough. Carbonate rock is dissolved while the site is below the water table. When the water table falls as an area is drained, the cavern is exposed. Little erosion takes place underground apart from solution action, which is concentrated along bedding planes, joints, faults, and other places of weakness.

Excess mineral matter carried in solution by groundwater may be deposited in open spaces. The most important such deposition occurs in the pores of sedimentary material, cementing it together to make sedimentary rock. Natural cement is most often calcium carbonate, iron oxide, or silica. The nature of the cement determines the strength and the color of most sedimentary rock. Rounded nodules (called concretions) and petrified wood are also formed in this way. As saturated groundwater percolates into a cavern now standing above the water table, it deposits dripstone as stalactites, stalagmites, and other forms.

19. Cavern formation is described most aptly as due to
    (1) deposition
    (2) erosion
    (3) gravity
    (4) petrification
    (5) solution

20. The process of cavern formation does *not* require
    (1) carbonate rock
    (2) carbon dioxide
    (3) drainage
    (4) silica
    (5) water

21. Buried wood may be petrified by the
    (1) addition of carbon dioxide
    (2) deposition of silica
    (3) formation of caverns
    (4) solution along bedding planes
    (5) solution of wood

22. Which of the following features has *not* formed by deposition?
    (1) concretion
    (2) dripstone
    (3) karst
    (4) stalagmite
    (5) cement in a sandstone

Growth movements are caused by unequal growth on opposite sides of organs. Plants exhibit two types of growth movements: induced movements (tropisms and nastic movements), which are influenced by external stimuli, and spontaneous movements (such as nutations), which are controlled by internal stimuli.

Tropisms are responses to any of several stimuli including light (phototropism), gravity (geotropism), touch (thigmotropism), and chemicals (chemotropism) that come primarily from one direction. Stems are positively phototropic and usually are negatively geotropic; they bend toward a bright light source and commonly away from the earth. Roots are either negatively phototropic or do not respond to light; they are positively geotropic. Tendrils of many climbing plants are positively thigmotropic, continuing to wind around an object with which they have made contact. The growth of a pollen tube toward an ovule is a positive chemotropism.

Nastic movements, like the daily rising and lowering of flower petals, are responses to changes in light or temperature, but they are independent of the direction from which the stimulus comes. Nutation is the spiral movement made by a stem tip as it grows in space.

23. Which type of movement is *not* controlled by environmental factors?
    (1) chemotropic
    (2) induced
    (3) nastic
    (4) nutation
    (5) thigmotropic

24. Which factor apparently fails to cause a tropism?
    (1) chemicals
    (2) gravity
    (3) light
    (4) temperature
    (5) touch

25. A stem that is *positively* phototropic and *positively* geotropic will bend
    (1) around an object
    (2) away from both the earth and the light source
    (3) away from the earth and toward the light source
    (4) away from the light source and toward the earth
    (5) toward both the earth and the light source

26. A vine winds along a fence especially in response to
    (1) chemicals
    (2) gravity
    (3) light
    (4) temperature
    (5) touch

A colloid is a suspension of small particles of one substance in another substance. Colloids show the following properties: Most colloids can pass through filter paper but not through semipermeable membranes. They may be purified by dialysis. Colloids scatter light so that the path of a light beam shone through a colloidal solution may be seen (Tyndall effect). Brownian movement (rapid jiggling) of colloidal particles can be seen microscopically. The surfaces of colloidal particles may selectively adsorb ions or molecules. In colloids, the boiling and freezing points of a liquid are little changed by the presence of the small, dispersed particles. Some colloidal particles are charged as shown by their movement toward an electrode (electrophoresis). Charged colloidal particles attract ions of opposite charge and mutual precipitation results.

Much particulate matter is treated in what industry calls Cottrell precipitators. The particles are charged or are made to adsorb an electric charge. If the air or gas containing these particles is passed between plates across which a high voltage is maintained, the charged particles will move to one of the plates. At intervals the current is turned off and the plates are shaken, allowing the accumulated particles to settle to a heap at the base of the precipitator for easy removal. Smelters make particular use of these precipitators to remove dust, which may contain valuable metals, from stack gases. Other industries employ Cottrell precipitators to prevent atmospheric pollution by particulate wastes.

27. As a rainstorm clears and the sun's rays are seen piercing the clouds, one sees
    (1) adsorption
    (2) Brownian movement
    (3) colloidal precipitation
    (4) electrophoresis
    (5) the Tyndall effect

28. Colloids are stable when their particles are charged and repel each other. So precipitation will occur when the colloidal particles are
    (1) adsorbed
    (2) boiled or frozen
    (3) dispersed
    (4) filtered
    (5) neutralized

29. The Cottrell precipitator is based on the phenomenon of
    (1) Brownian movement
    (2) dialysis
    (3) electrophoresis
    (4) smelting
    (5) the Tyndall effect

30. Which pollutant might a Cottrell precipitator suitably remove?
    (1) carbon monoxide from automobile exhaust
    (2) chlorine from a chemical plant
    (3) fine dust from a brickworks
    (4) hydrocarbons from an oil refinery
    (5) phosphates from residential sewage

ANSWER KEY FOR THE MINI-TEST

The Answer Key has been coded to assist you in identifying subjects for review.

        B—Biology
        G—Geology
        C—Chemistry
        P—Physics

1. (4) B
2. (2) C
3. (1) P
4. (5) C
5. (4) G
6. (5) B
7. (4) C
8. (5) P
9. (4) B
10. (2) G
11. (2) C
12. (5) C
13. (1) C
14. (5) C
15. (1) B
16. (3) B
17. (5) B
18. (1) B
19. (5) G
20. (4) G
21. (2) G
22. (3) G
23. (4) B
24. (4) B
25. (5) B
26. (5) B
27. (5) P
28. (5) P
29. (3) P
30. (3) P

SCORING YOUR GED SCIENCE TEST

To score your GED Science Test total the number of correct answers for the test. Do not subtract any points for questions attempted but missed, as there is no penalty for guessing. This score is scaled from 20 to 80. GED score requirements vary from state to state.

MINI-TEST: SCORE APPROXIMATOR

The following Score Approximator is designed to assist you in evaluating your skills, and to give you a very general indication of your scoring potential.

Number of Correct Answers	Approximate GED Score
25–30	62–80
20–24	53–61
15–19	46–52
10–14	39–45
5–9	24–38
1–4	20–23

ANALYZING YOUR TEST RESULTS

The charts on the following pages should be used to carefully analyze your results and spot your strengths and weaknesses. The complete process of analyzing each subject area and each individual question should be completed for each practice test. These results should be reexamined for trends in types of errors (repeated errors) or poor results in specific subject areas. THIS REEXAMINATION AND ANALYSIS IS OF TREMENDOUS IMPORTANCE TO YOU IN ASSURING MAXIMUM TEST PREPARATION BENEFIT.

MINI-TEST: GENERAL ANALYSIS SHEET

	Possible	Completed	Right	Wrong
Part A Background Knowledge	10			
Part B Reading Comprehension	20			
TOTAL	30			

MINI-TEST: SUBJECT AREA ANALYSIS SHEET

The coded Answer Key will assist you in this analysis.

	Possible	Completed	Right	Wrong
Biology	11			
Geology	6			
Chemistry	7			
Physics	6			
TOTAL	30			

ANALYSIS—TALLY SHEET FOR QUESTIONS MISSED

One of the most important parts of test preparation is analyzing WHY you missed a question so that you can reduce the number of mistakes. Now that you have taken the Mini-Test and corrected your answers, carefully tally your mistakes by marking them in the proper column.

	REASON FOR MISTAKE			
	Total Missed	**Simple Mistake**	**Misread Problem**	**Lack of Knowledge**
Biology				
Geology				
Chemistry				
Physics				
TOTAL				

Reviewing the above data should help you determine WHY you are missing certain questions. Now that you have pinpointed the type of error, when you take Practice Test 1 (following your study of the Subject Area Reviews) focus on avoiding your most common type.

COMPLETE ANSWERS AND EXPLANATIONS FOR THE MINI-TEST

Part A: General Science Knowledge

1. (4) The trachea is part of man's respiratory system, being the windpipe between the mouth and the lungs. The basic human circulatory system is heart to artery to capillary to vein; the aorta is the large artery from the left ventricle of the heart.

2. (2) Ammonia in solution is a base, ammonium hydroxide. Litmus paper is a pH *indicator,* turning blue in bases and red in acids. Two other well-known indicators are methyl red (for acids only) and phenolphthalein (for bases only).

3. (1) A light-year is the distance light travels in one year. Since the speed of light is an incredible 186,000 miles per second, a light-year is about six trillion miles. This is a convenient unit for describing the immensity of interstellar space. The star nearest to the sun is four light-years away.

4. (5) Iron rusts by combining with oxygen from the air. Combustion, decay, and respiration all involve the oxidation of organic (carbon-rich) matter, so these processes consume oxygen and produce CO_2. Photosynthesis in plants is the reverse, absorbing CO_2 and yielding oxygen.

5. (4) Seismographs are instruments that detect earthquakes, and the graphic record is called a seismogram. Barometers measure air pressure. A bathyscaphe is a vessel used for deep-sea diving. A spectrometer detects the presence and abundance of specific chemical elements from their characteristic optical spectra.

6. (5) All four animals have backbones and belong to the phylum of chordates. Amphibians are vertebrates that live both in water and on land; the frog is an example. Mammals are vertebrates that are warm-blooded and suckle their young with mammary glands; the elephant is a mammal. Carnivores are meat-eating mammals, including dogs and cats. Terrestrial organisms live on land, but the shark is marine.

7. (4) Gases react with simple ratios of their volumes. This law led John Dalton to speculate on the atomic nature of matter. The volumes (and therefore numbers of molecules) are proportional to the coefficients in an equation. In our reaction the coefficients are 2 for hydrogen and water and an implied 1 for oxygen. So 2 volumes of hydrogen react with 1 volume of oxygen to yield 2 volumes of water. The ratio of reactants:product is 3:2.

With 12 liters of water vapor, there must have reacted 18 liters of hydrogen plus oxygen.

8. (5) Since the experiment would be performed in a vacuum, there would be no air friction. Any material would accelerate at a uniform 32 ft/sec^2. The speed depends only on the distance of fall. Recall Galileo dropping large and small lead balls from the Leaning Tower of Pisa.

9. (4) Figure 1 shows a Mendelian table for the cross of two hybrids, each with orange (O) and yellow (Y) alleles. Each box represents an equal fraction of the offspring. 25% would be purebred orange (OO), and 50% would be hybrid orange (OY)—recall that O is dominant over Y—for a total of 75% orange marigolds. Only 25% would appear yellow (YY) because Y is recessive.

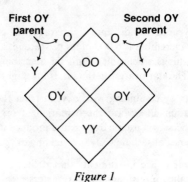

Figure 1

10. (2) Most large rivers have tributaries (smaller, contributing streams) and wind in tortuous meander bends across floodplains. A river usually deposits a triangular delta at its mouth. However, a moraine is a glacial deposit and has no necessary relation to a river.

Part B: Science Reading Comprehension

11. (2) The passage states that hydrogen escapes because of its low density. Density is mass divided by volume. Hydrogen has low density because it has low mass.

12. (5) The electrolysis of water would also yield oxygen:

$$H_2O \rightarrow H_2 + O_2$$

Sometimes the oxygen combines with the metal electrode to produce a metallic oxide.

13. (1) The commercial process has steam react with carbon to yield a gaseous mixture of carbon monoxide and hydrogen. All gases are miscible.

14. (5) The hydrogen comes from both the methane (CH_4) and steam (H_2O). The catalyst is a substance which speeds the reaction without itself being changed.

15. (1) Fermentation first proceeds aerobically (with air), then anaerobically (without it).

16. (3) Fermentation gradually converts the grape sugar to alcohol. Yeasts reproduce rapidly through fermentation.

17. (5) *Saccharomyces* is the generic name of the yeast for commercial wine.

18. (1) In home fermentation the grape juice has yeast introduced on the surface of the grapes as well as on grape leaves.

19. (5) Caverns are formed by the underground dissolving of carbonate rock. Erosion is not the best answer, as it is too general. Solution is the type of erosion with which we are concerned.

20. (4) Carbon dioxide and water combine to dissolve carbonates below the water table; with subsequent drainage of the area a hollow cavern is exposed. However, silica (SiO_2) is a common natural cement unrelated to cavern formation.

21. (2) Wood is petrified when subsurface water deposits a natural cement (silica) in the pores of the wood. The resulting silica rock retains the structure and appearance of the original wood. Choice (5) is misleading because wood does not dissolve but decay.

22. (3) Percolating solutions deposit cement or concretions in sediments. Other solutions deposit dripstone, including stalagmites and stalactites, in caverns. Karst, however, is terrain riddled with sinkholes from the solution of limestone.

23. (4) A nutation is a spontaneous movement which is controlled by internal stimuli. Tropisms and nastic movements are induced by external conditions.

24. (4) The first sentence of the second paragraph lists tropisms caused by light, gravity, touch, and chemicals. Temperature is only mentioned as inducing a nastic movement. The difference between a tropism and a nastic movement is that the latter are nondirectional.

25. (5) A positive tropism is toward the external stimulus, so the stem

would bend toward the light (positively phototropic) and toward the earth (positively geotropic). A negative tropism is away from the stimulus. Note that the passage states that most stems are negatively geotropic, bending away from the earth.

26. (5) Thigmotropism is growth induced by touch, like a vine climbing a fence.

27. (5) Moisture droplets in air would be colloidal. The Tyndall effect is where a light beam is visible through a colloidal solution.

28. (5) When charged colloidal particles are electrically neutralized by ions of opposite charge, mutual precipitation results.

29. (3) Electrophoresis is the movement of charged colloidal particles toward an electrode. Thus the Cottrell precipitator attracts particles out of the air (or other gas) to a charged plate.

30. (3) The Cottrell precipitator attracts particles from a gas and might prevent dust emission from a brick factory. Carbon monoxide and chlorine are gases, as are most of the hydrocarbons emitted from a refinery. Phosphates are not colloids but ions (PO_4^{-3}) in true solution.

PART IV: Subject Area Reviews

How to Review
Basic Concepts
Glossaries

HOW TO REVIEW

The following pages are designed to give you an intensive review of the major concepts used on the GED Science Test. The science reviews provide explanations of the main ideas in each field. Read them slowly and carefully, underlining key words and phrases. Do not spend excessive time on any one section of the review. If you are having difficulty understanding a concept, write a question mark (?) in the margin and return later. Further reading may clarify the concept.

A glossary of scientific terms associated with each of the major concepts in the test is also included. This is a valuable reference tool, since it provides definitions of terms that are used in the practice tests as well as the actual GED Science Test. Do not attempt to memorize these lists; rather, use the glossary together with the reviews to provide a mini-course approach to the science areas. When reviewing the terms think about which terms are related or contrasted, like *atom* and *molecule*. Write your own comments after many of the terms; make helpful notes.

Review an hour or two each day. Lengthy cram sessions are exhausting and ineffective. Many short review sessions provide the best preparation.

Reading the science section of a news magazine can be very helpful to gain familiarity with scientific terms and concepts. Topics of current interest are especially likely to appear on the GED.

BIOLOGY REVIEW

BASIC CONCEPTS

BIOLOGY is the science of life. Life has astonishing variety, embracing bacteria and baboons, whales and walnuts, algae and alligators—yet all those life forms share some similar materials and processes. The complexity of life compels biologists to specialize in certain levels of life: organic molecules, cells, organs, individuals, species, and communities. Here are some important characteristics of most life forms. A living organism has a very complicated *organization* in which a series of *processes* takes place. Life *responds* to its environment, often with *movement*. An organism must *maintain* itself and *grow*. Finally, a plant or animal will produce new organisms much like itself; *reproduction* is the most universal process of life, explaining its survival and variety.

THE CELL is the smallest amount of living matter, a bit of organic material that is the unit of structure and function for all organisms. Cells range in size from the smallest speck visible through an excellent microscope to the yolk of the largest egg. Some tiny organisms (like bacteria) are one-celled, but all larger organisms are composed of many cells arrayed in tissues. Although an isolated cell may be spherical, the cells packed together in plant or animal tissues have flattened walls. The essential subdivisions of a cell are the cell membrane, the cytoplasm, and the cell nucleus. (See figure 2.) The cell membrane is semipermeable, allowing some substances

THE CELL

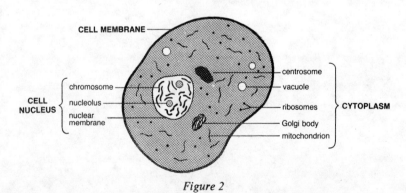

Figure 2

to pass while excluding others. The main material within a cell, the cytoplasm, varies in consistency from a fluid to a semisolid. Embedded in the cytoplasm are functional bodies: the centrosome which participates in cell division, ribosomes for constructing proteins, mitochondria which conduct metabolism, the Golgi bodies involved in secretion, and vacuoles used in digestion. The cytoplasm of plant cells also contains plastids, bodies with chlorophyll which carry out photosynthesis. The cell nucleus is a separate mass containing nucleoli and chromosomes, the genetic material.

METABOLISM is the set of chemical reactions within protoplasm, the living material of the cell. The chemical constituents of protoplasm include water as well as organic and inorganic compounds. The organic molecules are proteins, carbohydrates, lipids, and nucleic acids. Proteins are both structural components and enzymes, organic catalysts that enable particular metabolic reactions to proceed; all proteins are built from simpler amino acids. Carbohydrates (starches and sugars) and lipids (fats) are energy sources for cellular processes. The two nucleic acids, deoxyribose nucleic acid (DNA) and ribose nucleic acid (RNA), are complex chained molecules with encoded instructions for metabolism; the chromosomes of the cell nucleus contain the DNA. Metabolic reactions involve assimilation, photosynthesis, digestion, and respiration. The result is to store chemical energy as adenosine triphosphate (ATP). During cellular work the ATP decomposes and yields energy.

MITOSIS is the process of cell division in which the nuclear material of the original cell is divided equally between the newer cells. Such multiplication of cells permits growth of an organism, whether it occurs in the core of a tree or the muscle of a man. (See figure 3.) Mitosis begins as the chromosomes thicken and the centrosome divides (prophase). Then the nuclear membrane disappears and a spindle develops between the two parts of the centrosome. The chromosomes gather on that spindle (metaphase). The spindle divides, splitting each chromosome apart (anaphase). Finally, the nuclear membranes form and two new cells result (telophase). The chromosomes carry a genetic message enabling the cell to make proteins, so mitosis provides a complex mechanism for a new cell to obtain the genetic instructions it needs.

MITOSIS

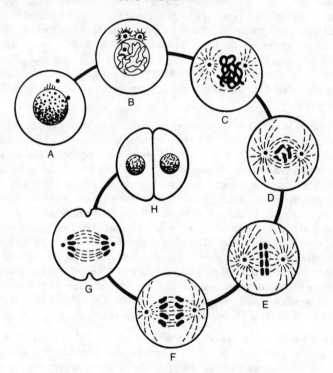

A, resting cell; B, C, and D, prophase;
E, metaphase; F, anaphase; G and H, telophase.

Figure 3

PLANTS may be divided into five broad groups. The more primitive groups are *algae* and *fungi;* these plants lack true roots, stems, and leaves. Algae range from a single cell to huge seaweeds; mostly they inhabit lakes and oceans. The fungi include molds, yeasts, and mushrooms. Fungi lack chlorophyll and thus are incapable of manufacturing food, so they are either parasites, preying on other living organisms, or saprophytes, existing on waste products and decaying organisms. A lichen is actually two organisms, a fungus and an alga, living together symbiotically. The more advanced plants possess roots, stems, and leaves. The *ferns* lack seeds and reproduce by means of spores, each of which may develop into a new plant

TABLE 1

Plant Group	Chlorophyll	Leaves	Seeds	Flowers
Fungi	no	no	no	no
Algae	yes	no	no	no
Ferns	yes	yes	no	no
Gymnosperms	yes	yes	yes	no
Angiosperms	yes	yes	yes	yes

without fertilization. Unlike the ferns, the seed plants require fertilization, and male pollen grains are carried to the female ovule by the wind. The *gymnosperms* are cone-bearing plants (including pines) with seeds exposed on cone scales. The *angiosperms* are flowering plants which bear their seeds within fruits. (See figure 4.)

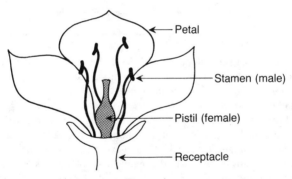

Figure 4

PHOTOSYNTHESIS is the process by which plants convert light into the chemical energy stored in foods. Chlorophyll, the green pigment within the leaf, is necessary to trap light energy for the photosynthetic reaction. In photosynthesis, carbon dioxide and water react to form sugar and oxygen:

$$6CO_2 + 6H_2O + \text{Light} \xrightarrow{chlorophyll} C_6H_{12}O_6 + 6O_2$$

carbon dioxide water energy $\qquad\qquad$ sugar oxygen

Thus plants absorb carbon dioxide and give off oxygen. The sugar made by plants can be oxidized later in a process that releases energy. The oxidation can be in the plant itself or in an animal that eats the plant. The release of energy by oxidation of sugar is respiration:

$$\underset{sugar}{C_6H_{12}O_6} + \underset{oxygen}{6O_2} \xrightarrow{Krebs\ cycle} \underset{\substack{carbon \\ dioxide}}{6CO_2} + \underset{water}{6H_2O} + Energy$$

TABLE 2

Photosynthesis	Respiration
Carbon dioxide and water are raw materials	Sugar and oxygen are raw materials
Sugar and oxygen are produced	Carbon dioxide and water are produced
Reduction reaction	Oxidation reaction
Occurs only in light	Occurs in light or darkness
Occurs only in chlorophyll-containing cells	Occurs in all living cells
Occurs in chloroplasts	Occurs in mitochondria

ANIMALS cannot perform photosynthesis and, therefore, derive their food from other organisms. Herbivores eat plants directly. Carnivores prey on other animals, but this food chain, too, ends in plants. Plants and animals are classified into *phyla* on the basis of their cells, tissues, organs, and overall organization. Each phylum is a major group of organisms. There are about 6 phyla of protists (one-celled organisms), 8 phyla of plants, and 21 phyla of animals. Table 3 lists the most important animal phyla. For the last three, advanced phyla, the main classes of animals are also given. There are some small marine chordates of primitive form, but those chordates listed are the vertebrates, characterized by a backbone, a definite head, well-developed brain and eyes, a central heart, red blood cells, and two pairs of limbs. Let's take a closer look at the organs within the vertebrates, using man as our example.

TABLE 3

Phylum	Members and Description
Porifera	Sponges; ingest microscopic food from currents
Coelenterata	Corals; capture food with stinging tentacles
Platyhelminthes	Flatworms, tapeworms; most are parasitic
Nematoda	Roundworms, hookworms; some are parasitic
Annelida	Segmented worms, including earthworms
Bryozoa	Colonial animals in ocean
Brachiopoda	Lampshells; bivalves with tentacles inside shell
Echinodermata	Starfish, sea urchins, crinoids; 5-rayed symmetry
Mollusca	Pelecypoda: clam, oyster, mussel; bivalves Gastropoda: snail; often with coiled shell Cephalopoda: octopus, squid
Arthropoda	Crustacea: crab, crayfish, lobster Insecta: ant, bee, fly, grasshopper; 6 legs Arachnida: spider, tick, scorpion; 8 legs
Chordata	Agnatha: lamprey; primitive fishes with round mouths Chondrichthyes: shark, ray; skeleton of cartilage Osteichthyes: cod, perch, trout; bony fishes Amphibia: frog, toad, salamander; lay eggs in water Reptilia: snake, turtle, alligator; lay eggs on land Aves: sparrow, pigeon, chicken; birds with feathers Mammalia: cat, man, whale; females have mammary glands for milk

THE SENSORY SYSTEM includes those specialized structures which initiate a nerve impulse after being affected by the environment. The eyes are the organs of vision. Light rays are refracted as they pass through the cornea, lens, and vitreous body to focus on the retina, where an image is formed. The optic nerve then carries impulses from the light-sensitive cells of the retina to the brain. (See figure 5.) The ear is the receptor of sound and the organ of balance. When sound waves cause the tympanum (eardrum) to vibrate, three tiny bones transmit the motion to the cochlea and the organ of Corti, which initiates a nerve impulse. The seat of balance is in the semicircular canals within the inner ear. The receptors of taste are distributed on the upper surface of the tongue. Each taste bud detects one of the four primary tastes (salty, sweet, sour, and bitter); other taste sensations are combinations of the primary tastes.

THE EYE

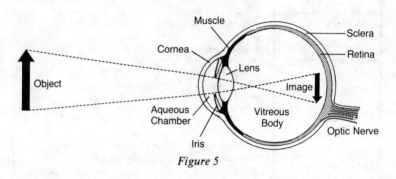

Figure 5

THE NERVOUS SYSTEM is composed of the brain, spinal cord, and peripheral nerves which extend throughout the body. The functional unit of the nervous system is the neuron, a nerve cell with short dendrites that carry electrical impulses to the cell body, and a long axon, the outgoing fiber along which the impulse is transmitted further. (See figure 6.) Sensory neurons conduct signals from sense organs to the central nervous system, the spinal cord, and brain. Motor neurons transmit signals from the central nervous system to muscles. Figure 7 shows several major parts of the human brain. The hindbrain (cerebellum and medulla oblongata) operates unconsciously and automatically to regulate vital functions like circulation, respiration, excretion, and muscle tension. The cerebrum is the largest part of the brain; it receives information from the senses and makes conscious decisions.

NEURON

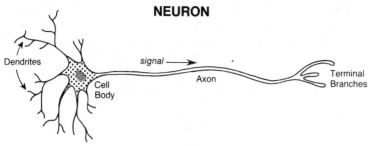

Figure 6

THE BRAIN

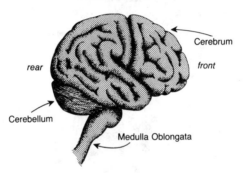

Figure 7

THE DIGESTIVE SYSTEM includes the mouth, pharynx (throat), esophagus, stomach, small intestine, and large intestine. Other organs in this system are the salivary glands, liver, gallbladder, and pancreas. The enzymes contained in saliva, gastric juice, pancreatic juice, and intestinal fluids convert carbohydrates, fats, and proteins into molecules small enough to be absorbed into the blood. Simple sugars are absorbed as such and do not require digestion. Carbohydrates are converted to various sugars by the action of several enzymes, including ptyalin from saliva. Fats are transformed to glycerol and fatty acids by the combined action of bile from the liver and the enzyme lipase from the pancreas. Proteins are broken apart to their constituent amino acids. The final products of digestion—sugars, glycerol, fatty acids, and amino acids—are absorbed into the bloodstream through the millions of projections (villi) lining the small intestine. Once in the blood, these molecules are metabolized in the various body tissues.

THE CIRCULATORY SYSTEM consists of the blood, the heart, and the blood vessels. (See figure 8.) The blood is composed of red cells, white cells, and platelets suspended in a watery medium called plasma. Red cells transport oxygen in combination with the iron pigment, hemoglobin.

THE CIRCULATORY SYSTEM

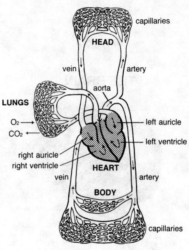

Figure 8

Human hemoglobin is a protein with the formula $C_{3032}H_{4816}O_{872}N_{780}S_8Fe_4$ (C stands for carbon, H hydrogen, O oxygen, N nitrogen, S sulfur, and Fe iron) and that giant molecule transports just 4 oxygen molecules. The function of white blood cells is to fight infection, while platelets initiate the clotting necessary to stop bleeding after a wound. Nutrients, wastes, hormones, antibodies, and enzymes are dissolved in the plasma. The heart is a muscular pump, beating about 70 times each minute. Blood flows from the heart through arteries and returns to the heart in the veins. The pulmonary circulation is the flow of blood between the right side of the heart and the lungs, where oxygen diffuses into the blood and carbon dioxide leaves the blood. Remember that respiration enables the body to oxidize food, consuming oxygen and producing carbon dioxide. From the lungs, the oxygenated blood returns to the left side of the heart, where the powerful left ventricle pumps it through the aorta into general circulation. Arteries lead into smaller vessels called capillaries, where oxygen and nutrients diffuse into tissue cells. Wastes diffuse back into the capillaries, which lead into veins and, ultimately, back to the right side of the heart.

REPRODUCTION in organisms may occur by either sexual or asexual processes. In asexual reproduction there is only one parent and simple mitotic division produces the offspring; most protists, many plants, and a few primitive animals follow this reproductive strategy. Sexual reproduction involves two parents, as in all higher animals, except in the cases of self-fertilization in many flowering plants. The advanced plants and animals produce male and female sex cells by a special mode of cell division called meiosis. The normal adult organism has a double set of chromosomes within each cell, a condition described as diploid. A haploid cell has only half the normal number of chromosomes. Figure 9 shows the four pairs of chromosomes within the diploid cells of the fruit fly, *Drosophila*. The meiotic division employs the same spindle machinery as used in mitosis, but the consequence is not diploid daughter cells identical to the parent (as in mitosis) but haploid sex cells, the gametes. Gametes from a male are sperm cells, while gametes from a female are egg cells (ova). A haploid sperm must unite with a haploid egg to form a diploid zygote, which develops into a mature organism through the process of embryogeny. Meiotic production of gametes provides the means of mixing genetic information through the sexually reproducing population, and the spread of useful information allows successful evolutionary adaptation.

MEIOSIS

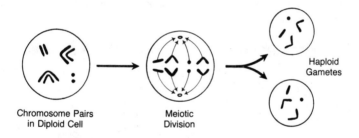

(cell components besides chromosomes not shown)

Figure 9

NUCLEIC ACIDS store genetic messages and instruct the cell how to make the many proteins needed for life. Heredity has an elegant molecular basis. Nucleic acids are truly the secret of life. DNA in the chromosomes of the cell nucleus is a long molecule of two chains twisted into a double spiral or helix. Nitrogenous bases are attached to the chains in a special sequence that is a coded message, the instructions for life. The double chain permits the DNA to duplicate itself exactly during reproduction, retaining the original genetic message. The message contains the instructions necessary to build various proteins, many of which function as enzymes. Some RNA molecules carry the message from the DNA in the cell nucleus to the ribosomes in the cytoplasm, where protein assembly occurs.

GENES are the functional message units along the DNA chains within the chromosomes. One chromosome is composed of many genes, each of which determines or influences an inheritable trait. In diploid organisms, chromosomes occur in pairs and so must genes. The first man to study such pairs of genetic traits was the Austrian monk Gregor Mendel (1865) who cultivated garden peas, the seeds of which were either round or wrinkled. He discovered that when he crossed purebred round peas with wrinkled peas, the first hybrid generation was all round peas. Mendel inferred that the round characteristic (R) was dominant over the wrinkled characteristic (w) in the hybrid plants (Rw). In figure 10, each adult pea plant has two genetic characteristics (alleles), while the gametes have only one allele apiece. Then the hybrid peas were crossed and the next generation was 75% round and 25% wrinkled. (See figure 11.) Notice that the recessive wrinkled trait can only manifest itself in a plant with two wrinkled alleles (ww), but round pea seeds occur with either (RR) or (Rw). Mendelian genetics permits understanding of the frequencies of various genetic traits in populations. New alleles arise by mutations, errors during copying of the DNA code.

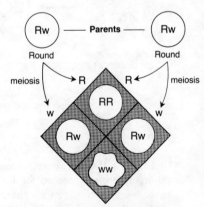

Figure 10   *Figure 11*

EVOLUTION of life is indicated by the fact that fossil organisms in rock strata are different from modern organisms. As we go back in time, searching lower and older strata, the organisms diverge more and more from those living today. Yet the variation in life forms appears to be relatively continuous. For example, 60 million years ago horses were quite small and had four toes on each foot. As time passed, horses evolved through a series of larger sizes and fewer toes to today's large, single-toed creature. Other evidence for the evolution of life comes from the study of biogeography (the distribution of present-day species), embryology (the similarities among early developmental stages of animals), homology (structural similarities in various organisms), and biochemistry (chemical similarities in various organisms). Charles Darwin listed evidences for the progression of life in his book *The Origin of Species* (1859) and proposed that evolution proceeded by natural selection. Within a variable population those organisms able to adapt to the environment most successfully would tend to have more offspring than their less successful rivals, and so the genetic characteristics of the entire population would slowly shift toward better adaptation. In this manner, given the immensity of geologic time, entire new organic structures, systems, and species could arise. Modern evolutionary biologists emphasize the central role of mutation in providing the genetic variation necessary for nature's trial-and-error selection.

TAXONOMY is the classification and naming of organisms. Over 1,300,000 different species have been described, so it is essential to sort them systematically. The binomial nomenclature devised by the Swede Linnaeus (1735) gives each organism two names, the genus and species. For example, the dog is *Canis familiaris*. Genera are grouped into higher taxonomic levels, arranged in a hierarchy as shown in table 4. At the

TABLE 4

Kingdom
Phylum
Class
Order
Family
Genus
Species

highest level are the three kingdoms: protists, plants, and animals. Organisms are classified into taxonomic groups by morphological similarities and genetic affinities.

ECOLOGY is the study of the relationship of organisms to their environment. The major habitats are the oceans, fresh water, and land. The oceans have the greatest proportion of living things. The upper layers of the oceans contain microscopic plants collectively called phytoplankton. Through photosynthesis, phytoplankton produce food for the marine life of the depths. Because ocean conditions are relatively uniform, most marine species are broadly distributed. Rivers, brooks, lakes, and swamps house different species of organisms. Fresh water shows greater variation in currents, composition, and temperature than seawater, so many freshwater species are quite restricted in distribution. Land habitats show the greatest extremes in temperature and moisture, with a corresponding diversity in life. In any habitat the various organisms compete for food. The ultimate source of food is photosynthetic plant life.

GLOSSARY OF TERMS IN BIOLOGY

ADENOSINE TRIPHOSPHATE: A compound with energy-rich phosphate bonds involved in the transfer of energy in cellular metabolism; abbreviated ATP.
ADRENALIN: A hormone secreted by the adrenal medulla; also called epinephrine.
AGNATHS: A class of vertebrates without jaws; includes primitive fish like the lamprey and hagfish.
ALGAE: Simple plants containing chlorophyll.
ALIMENTARY CANAL: Organs of digestion through which food passes.
ALLELES: Alternative genes which produce contrasting characteristics.
AMINO ACID: An organic compound containing an amino and a carboxyl group; the building blocks of proteins.
AMPHIBIANS: Class of vertebrates capable of living both in water and on land. The larval forms have gills and the adults have lungs; includes frogs and toads.
ANGIOSPERMS: The class of flowering plants, with seeds enclosed in fruits.
ANNELIDS: The phylum of round, segmented worms.
ANTERIOR: Toward the forward end.
ANTHER: The part of a flower that produces pollen.
ANTIBIOTIC: A substance that destroys a microorganism or inhibits its growth.
ANTIBODY: A substance produced by the body to combat the injurious effect of a foreign substance (antigen).
ANTISEPTIC: A substance which kills bacteria.
AORTA: The main artery leaving the heart.
ARACHNIDS: A class of arthropods with no antennae and four pairs of legs; includes spiders, scorpions, ticks, mites, and king crabs.

ARTERY: A blood vessel which carries blood away from the heart.
ARTHROPODS: The phylum of segmented invertebrates with jointed appendages and a chitinous exoskeleton. Includes arachnids, crustaceans, and insects.
ASEXUAL: Reproduction in one individual, without the union of gametes.
AURICLE: A chamber through which blood enters the heart; also called atrium.
AVES: A class of vertebrates with feathers and wings; the birds.
AXON: The nerve fiber that conducts an impulse away from the body of a nerve cell.
BACTERIA: Unicellular organisms without a distinct nucleus and usually without chlorophyll.

BACTERIA

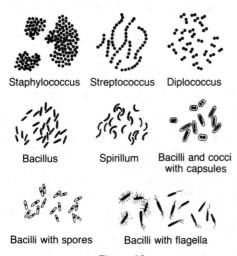

Figure 12

BILE: A yellowish-green fluid secreted by the liver; aids in the digestion of fats.
BINOMIAL NOMENCLATURE: The international system of naming organisms with two names, the first generic and the second specific.
BOTANY: The science of plants.
BRYOPHYTES: The phylum of mosses and liverworts.
BUDDING: Asexual reproduction by splitting of a new organism from the parent.
CAPILLARY: The smallest blood vessel which carries blood between an artery and a vein.

CARBON CYCLE: The exchange of carbon between living things and their environment.
CARNIVORE: A flesh-eating animal.
CARTILAGE: A firm connective tissue, more flexible than bone.
CELL: The basic unit of organic structure and function.
CELLULOSE: The woody tissue of plants.
CENTRAL NERVOUS SYSTEM: The brain and spinal cord.
CEREBELLUM: The part of the vertebrate brain which controls muscular coordination.
CEREBRUM: The upper part of the brain, where conscious mental processes occur.
CHLOROPHYLL: The green coloring matter in a plant; facilitates photosynthesis.
CHONDRICHTHYES: A class of vertebrates comprising the cartilaginous fishes; includes sharks, rays, skates, and sawfishes.
CHORDATES: The phylum characterized by a spinal cord; includes the vertebrates.
CHROMOSOME: A body in the cell nucleus; the bearer of genetic information.
CLASS: The main subdivision of a phylum.
COELENTERATES: The phylum of corals and jellyfish.
COLD-BLOODED: An animal whose body temperature varies with the surroundings.
COLONY: A group of individuals of one species living together.
COMMUNITY: A group of individuals of many species living together.
CONIFER: A cone-bearing tree.
CORAL: A colonial marine coelenterate.
COROLLA: The petals of a flower.
CORPUSCLE: A blood cell.
CROSS-POLLINATION: The transfer of pollen from one plant to a flower on another plant.
CRUSTACEANS: The class of arthropods with gills and two pairs of antennae; includes lobsters, crabs, barnacles, and crayfish.
CYTOPLASM: The substance of the cell outside the nucleus.
DENDRITE: The branching nerve fiber that conducts impulses toward the body of a nerve cell.
DEOXYRIBOSE NUCLEIC ACID: The compound in the chromosomes that stores genetic information as a molecular code; abbreviated DNA.
DERMIS: The inner layer of the skin.
DIATOMS: Algae of planktonic habit and with siliceous cases.
DIGESTION: The breakdown of food for absorption.
DIPLOID: Having chromosomes in homologous pairs.
DOMINANT: The one of two alternative genetic traits which is displayed in a heterozygous individual.
DORSAL: Toward the upper side; the back.

ECHINODERMS: The phylum of radially symmetrical marine animals with spiny exoskeletons; includes the starfish, sand dollar, and sea urchin.
ECOLOGY: The study of relations between organisms and their environment.
EMBRYO: An organism in the early stages of development.
ENDOCRINE GLAND: A gland which secretes a hormone.
ENVIRONMENT: The conditions in which an organism lives.
ENZYME: A protein that serves as an organic catalyst for metabolic reactions.
EPIDERMIS: The outer layer of the skin.
EVOLUTION: The modification of life forms with the passage of time.
EXCRETION: The discharge of waste materials.
EXOSKELETON: A hard, jointed case outside the fleshy tissues of an animal.
FAMILY: The main subdivision of an order.
FAUNA: The group of animals in an area.
FERTILIZATION: The union of gametes to form a zygote.
FLORA: The group of plants in an area.
FOSSIL: Any naturally preserved remains of ancient life.
FRUIT: The mature ovary of a flower.
FUNGI: Plants that lack chlorophyll; molds, mushrooms, and yeasts.
GAMETE: A sex cell; an egg or sperm.
GENE: A unit of heredity located on the chromosome.
GENETICS: The study of inheritable characteristics.
GENUS: The main division of a family.
GERMINATION: The sprouting of a seed.
GONAD: A gamete-producing organ in animals; testis or ovary.
GYMNOSPERMS: A class of vascular plants bearing seeds in cones.
HAPLOID: Having only half the normal number of chromosomes.
HEMOGLOBIN: The iron-bearing pigment of the red blood cells.
HERBIVORE: A plant-eating animal.
HEREDITY: The transmission of characteristics from parents to offspring.
HETEROZYGOUS: An individual that has two different genes for one particular character, like the (Rw) hybrid peas grown by Mendel.
HOMOLOGY: The similarity of body structures of different organisms, due to common ancestry; the structures may not have the same function. A bat's wing is homologous to a squirrel's foreleg.
HOMOZYGOUS: An individual with identical genes for one particular character, like the (RR) purebred peas studied by Mendel.
HORMONE: A chemical substance that regulates body processes.
HYBRID: The offspring of genetically different parents.
INSECTS: A class of arthropods characterized by 3 body sections, 6 legs, and usually 4 wings.
INSULIN: A hormone produced by the pancreas, which regulates the body's utilization of sugar.

LEGUMES: Plants (like beans and peas) whose roots have colonies of bacteria that change atmospheric nitrogen into a form usable by the plant.

LIGAMENT: A fibrous band which supports an organ or connects two bones.

LYMPH: A colorless fluid in the body surrounding many cells.

MAMMALS: A class of warm-blooded vertebrates possessing hair and feeding their young by means of mammary glands with milk.

MEDULLA OBLONGATA: The posterior part of the vertebrate brain.

MEIOSIS: The mode of cell division that produces gametes, each with one-half the number of chromosomes of the parent cell.

METABOLISM: The chemical processes within an organism.

METAMORPHOSIS: The change from a larval form to an adult form.

METAZOA: Multicellular animals.

MITOCHONDRIA: Bodies in the cytoplasm, containing enzymes.

MITOSIS: Cell division with chromosome duplication, forming offspring cells with the same number of chromosomes as the parent cell; cell-splitting.

MOLLUSKS: The phylum containing pelecypods (clams, oysters), gastropods (snails), and cephalopods (octopi, squids).

MUTATION: An inheritable change in a gene.

NATURAL SELECTION: The survival of the best-adapted organisms.

NERVOUS SYSTEM: The brain, spinal cord, and nerves.

NEURON: A nerve cell.

NUCLEUS: The central part of a cell, containing the chromosomes and controlling cellular activities.

OOGENESIS: The maturation of egg cells.

ORDER: The main division of a class.

ORGAN: A group of cells or tissues functioning as a whole.

ORGANISM: A living plant or animal.

OSTEICHTHYES: The class of vertebrates comprising the bony fishes.

OVARY: The part of an organism where eggs or seeds form.

OVUM: An egg; a female gamete.

PARASITE: An organism that lives in or on another organism, deriving food at the expense of its host.

PASTEURIZATION: The killing of microorganisms in milk by heating to 145°F for 30 minutes.

PENICILLIN: An antibiotic drug obtained from molds.

PESTICIDE: A substance used to destroy plants or animals.

PHALANGES: The bones of the toes and fingers.

PHENOTYPE: Appearance of an organism, as opposed to its genetic constitution.

PHOTOSYNTHESIS: The production of carbohydrates by green plants in the presence of light.

PHYLUM: A major group of animals or plants; the main division of a kingdom.
PISTIL: The central portion of a flower, consisting of the ovary, style, and stigma.
PITUITARY: The endocrine gland located at the base of the brain, whose hormones regulate other glands.
PLANKTON: The microorganisms that live in the ocean.
PLASMA: The liquid part of the blood.
PLASTIDS: Bodies in the cytoplasm of plant cells, involved in food synthesis.
PLATELET: A particle in the blood which promotes clotting.
PLATYHELMINTHES: An animal phylum containing the flatworms.
POLLEN: The mature microspores of seed plants.
POLLINATION: Fertilization by the transfer of pollen from an anther to a stigma.
PORIFERA: The animal phylum containing the sponges.
POSTERIOR: Toward the hind end.
PREDATOR: An animal that lives by preying on other animals.
PROTEIN: An organic compound made up of amino acids.
PROTISTA: The kingdom of one-celled organisms.
PROTOPLASM: A general term for the living matter of the cell.
RECESSIVE: The one of two alternative genetic traits which is masked in a heterozygous individual.
REFLEX. A response to a stimulus.
REPTILES: The class of scaly vertebrates which includes the snakes, turtles, lizards, alligators, and crocodiles.
RESPIRATION: Biological oxidation.
RIBOSE NUCLEIC ACID: A substance in the cell with the function of making proteins; abbreviated RNA.
RIBOSOMES: Bodies in the cytoplasm concerned with protein synthesis.
SEPTUM: A wall separating two cavities.
SEXUAL: Reproduction involving the union of an egg and sperm.
SPECIES: A group of similar animals or plants, usually capable of interbreeding; the main division of a genus.
SPERM: A male gamete.
SPORE: An asexual reproductive cell found in fungi and ferns.
STAMEN: The organ of a flower that produces pollen.
SYMBIOSIS: The close living association of organisms of different species in which both benefit.
TAXONOMY: The classification of organisms.
TESTIS: A male gonad producing sperm cells.
TISSUE: A group of cells having the same structure and function.
TOXIN: A substance produced by an organism that is poisonous to another organism.

TRAIT: An inherited characteristic.
TRANSPIRATION: The evaporation of water from plants.
TROPISM: A growth movement in a plant in response to an environmental stimulus.
VACCINE: A fluid containing dead disease germs injected into an animal to produce immunity.
VACUOLE: A fluid-filled space in the cytoplasm that contains food or wastes.
VEIN: A vessel conveying blood toward the heart.
VENTRAL: Toward the lower side.
VENTRICLE: A chamber from which blood leaves the heart.
VERTEBRATES: Chordates characterized by a well-developed brain, a backbone, and usually two pairs of limbs; includes the fishes, amphibians, reptiles, birds, and mammals.
VILLI: Tiny projections in the small intestine, through which digested food is absorbed.
VIRUS: A simple form of matter, on the borderline between inorganic chemicals and life; often infects higher organisms.
VITAMIN: An organic compound needed in small quantities for normal metabolism.
WARM-BLOODED: An animal with a constant body temperature.
XYLEM: The woody tissue of plants that carries fluids up to the leaves.

WOODY STEM

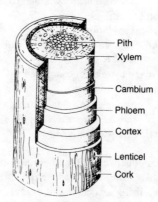

Figure 13

YEAST: A fungus that causes fermentation; used in baking and brewing.
ZOOLOGY: The science of animals.
ZYGOTE: The diploid cell resulting from the union of gametes; the fertilized ovum.

GEOLOGY REVIEW

BASIC CONCEPTS

GEOLOGY is the science that describes and interprets the earth. It classifies the materials that make up the earth, observes their shapes and distribution, and tries to discover the processes that caused the materials to be formed in that manner. Some major geological fields are geomorphology (landforms), petrology (rocks), stratigraphy (layered rocks), and paleontology (fossils). All fields contribute to historical geology, the ambitious attempt to list the specific events which have produced the present earth. Processes occurring today are observed carefully and their effects are measured. Then geologists assume that similar effects in ancient rocks were caused by processes similar to those of the present. This method of using the present to interpret the past is called uniformitarianism. For example, glaciation in early eras is indicated by ancient deposits with features very similar to those produced by present-day glaciers.

THE EARTH'S STRUCTURE (see figure 14) has been inferred from its astronomical properties and seismic records of earthquake waves which

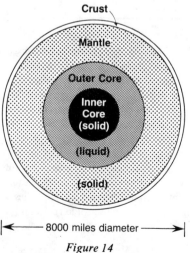

Figure 14

traveled through the interior of the earth. Temperature rises from the surface (20°C) to the center (3000°C) of the earth; this fact is essential to understanding geological processes. About 31% of the earth's mass is a dense core of iron and nickel metals, melted by the extremely high temperature of the center of the earth. Around that liquid core is the largest zone of the planet (68%), the mantle of crystalline silicates rich in magnesium, calcium, and iron. The very hot mantle is mainly solid, but local melting to magma (molten rock) is the source of volcanic eruptions. Above the mantle is the crust, which makes up less than 1% of the earth. This relatively thin zone (5–25 miles) contains the only rocks we can study, even in the deepest mines or drillholes. Table 5 shows the average chemical composition of crustal rocks.

TABLE 5
AVERAGE COMPOSITION OF CRUSTAL ROCKS

Element		Percent
Oxygen	O	62.6
Silicon	Si	21.2
Aluminum	Al	6.5
Sodium	Na	2.6
Calcium	Ca	1.9
Iron	Fe	1.9
Magnesium	Mg	1.8
Potassium	K	1.4

THE ROCK CYCLE (see figure 15) displays the linkage of processes within the crust. The earth's internal heat from radioactivity fuses (melts) solid rock to liquid magma, which has a temperature of about 1000°C. The magma is of lower density than the overlying rocks, so it tends to squeeze its way upward. Plutonic rocks form when the magma cools and crystallizes beneath the surface, but volcanic rocks form when the magma erupts at the surface. Exposure of any rock type at the surface leads to fragmentation by weathering and erosion of the particles. Sand and mud are deposited as soft sediments, which may be compacted by burial or cemented by subsurface solutions to form sedimentary rocks. Deeper burial of sedimentary and volcanic rocks to zones of higher temperature forces recrystallization to metamorphic rocks and, eventually, melting to magma.

GEOLOGY REVIEW 59

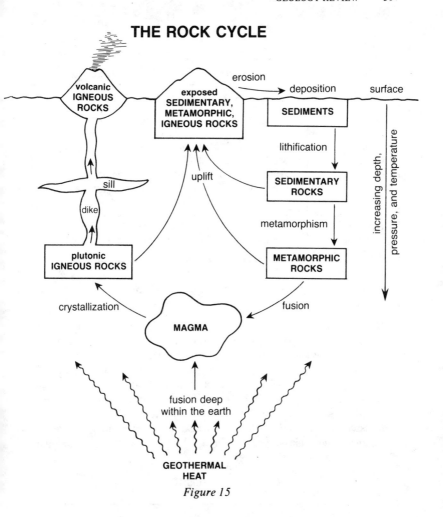

Figure 15

MINERALS are natural chemical compounds which are the crystals that make up rocks. Each mineral has a specific composition or narrow range of composition. In table 6 showing common rock-forming minerals, two chemical elements which may substitute for each other are enclosed by parentheses; thus the mineral olivine—$(Mg,Fe)_2SiO_4$—varies in composition from Mg_2SiO_4 to Fe_2SiO_4. The most abundant minerals in the crust are the two feldspars (orthoclase and plagioclase), quartz, olivine, and

TABLE 6

Mineral	Composition	Occurrence
Quartz	SiO_2	igneous, sedimentary, metamorphic
Orthoclase	$KAlSi_3O_8$	igneous, sedimentary, metamorphic
Plagioclase	$(Ca,Na)Al_2Si_2O_8$	igneous
Olivine	$(Mg,Fe)_2SiO_4$	igneous
Augite	$Ca(Mg,Fe)Si_2O_6$	igneous, metamorphic
Kaolin	$Al_2Si_2O_9H_4$	sedimentary: a clay mineral
Calcite	$CaCO_3$	sedimentary, metamorphic
Halite	$NaCl$	sedimentary: common salt
Hematite	Fe_2O_3	iron ore deposit
Chalcopyrite	$CuFeS_2$	copper ore deposit

augite. Note that these five minerals are silicates, built from interlocking silicon and oxygen atoms.

IGNEOUS ROCKS form by the cooling and solidification of molten rock matter. Magma that crystallizes within the earth in large volumes as batholiths cools slowly enough for large crystals to grow, so these plutonic (intrusive) rocks are recognized by their coarse grain size. Magma ascending toward the surface may follow sedimentary beds as sills or cut across the strata as dikes. Magma reaching the surface as a volcano may spill out as lava flows or explode into fragments which settle as ash deposits. Such volcanic (extrusive) rocks show quick chilling of lava and are fine-grained, glassy, or fragmental. Obsidian and pumice are familiar

TABLE 7
Common Igneous Rocks

Occurrence	Rock Name	Grain Size	Minerals
Plutonic	GABBRO	coarse	plagioclase, augite, olivine
	GRANITE	coarse	quartz, orthoclase, plagioclase
Volcanic	BASALT	fine	plagioclase, augite, olivine
	RHYOLITE	fine	quartz, orthoclase, plagioclase

glassy rocks. Notice that basalt is the volcanic equivalent of gabbro, and rhyolite is the volcanic equivalent of granite.

SEDIMENTARY ROCKS form by deposition at the earth's surface, usually in a body of water. Clastic sediments are accumulations of mud, sand, and pebbles from erosion of preexisting rocks. Chemical sediments are salts precipitated from seawater or a salt lake. Organic sediments were formed by life, usually by an aquatic creature extracting material from the water to form its shell; millions of shells may form a large bed or reef.

TABLE 8
COMMON SEDIMENTARY ROCKS

Process	*Rock Name*	*Description*	*Minerals*
Clastic	CONGLOMERATE	cemented pebbles	rock fragments
	SANDSTONE	cemented sand	quartz and orthoclase
	SHALE	hardened mud	clays, like kaolin
Chemical	SALT	crystalline	halite
Organic	LIMESTONE	shelly	calcite

METAMORPHIC ROCKS are produced when sedimentary or igneous rocks are transformed by high temperature or pressure. The extreme conditions force chemical reactions and recrystallization to new minerals and textures. Texture describes the size, shape, and orientation of mineral grains in a rock. Contact metamorphism is baking adjacent to an igneous intrusion. Regional metamorphism occurred where an entire area was buried and deformed during mountain building; such rocks have crystals with parallel orientations from the pressure of deformation.

TABLE 9
COMMON METAMORPHIC ROCKS

Rock Name	*Texture*	*Minerals*	*Parent Rock*
GNEISS	coarse oriented	quartz, orthoclase	shale, sandstone
SCHIST	medium oriented	quartz, micas	shale, sandstone
SLATE	fine oriented	micas	shale
MARBLE	coarse granular	calcite	limestone
QUARTZITE	medium granular	quartz	sandstone

WEATHERING is the destruction of bedrock by atmospheric action, with the generation of soil and loose, erodible debris. Most weathering involves chemical action by the air and water, which attack the minerals of the rocks. Iron-bearing minerals are oxidized, while common feldspar grains undergo hydrolysis to kaolin and other clays. Carbon dioxide in aqueous solution is a weak acid capable of slowly dissolving carbonates, like limestone; such limestone solution can yield a karst terrain pitted with small lakes and caverns. Weathering may involve only mechanical action, as in the shattering of rock by alternate freezing and thawing. The soil formed atop bedrock tends toward a profile of several layers (see figure 16), which may be inspected along highway roadcuts.

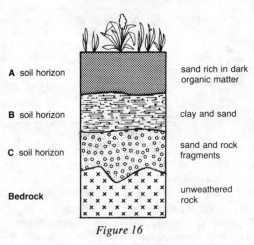

Figure 16

EROSION carries away the debris from weathering, moving it downslope to a more stable site of deposition. The most important erosional agent is running water. The water carries some salts in solution while particles are transported in suspension. Stream erosion of an uplifted highland leads to a characteristic time-series of landforms, from youthful terrain with V-shaped valleys to mature terrain with rounded hills and broad valleys. Wind is a significant means of erosion in arid regions; fine particles are blown away to be deposited downwind as sand dunes. At high elevations and polar latitudes ice is present throughout the year, so glacial erosion may occur. Snow accumulates and its weight transforms it to ice, which oozes slowly downslope under the pressure of its own weight. Glacial ice is a remarkably effective agent of erosion, capable of carving out U-shaped valleys in mountains. During the Pleistocene Ice Age, which ended only about 12,000 years ago, great glacial sheets covered much of North America and deposited an irregular blanket of till (mud and boulder debris).

STRATA are the layers of sediment deposited in a quiet environment. Common sites of deposition are lakes, deltas at the mouths of rivers, beaches and sandbars along the coast, and (most importantly) the marine environment. Strata are commonly very extensive laterally and relatively thin vertically, like a blanket. An important geological rule is the law of original horizontality, which states that most sediments are deposited in beds which were originally horizontal, and any tilting is due to later earth movements. A second stratigraphic principle is the law of superposition: younger beds were originally deposited above older beds.

FOSSILS are traces of ancient life preserved in the strata as shells, footprints, and the like. Because life has evolved (changed) continually through geological history, the fossils in older strata differ from those found in more recent deposits. In fact, strata deposited during one geological period contain characteristic life forms different from those of any other period. For example, the earliest fossil-rich beds have many trilobites, early crablike creatures which have been extinct for hundreds of millions of years; discovery of fossil trilobites in a formation permits assignment of that bed to an early period.

THE GEOLOGICAL TIME SCALE was a major achievement of stratigraphers, who used fossils to arrange strata in a standard order. More recently, geochemists have measured the amount of radioactive decay in minerals and calculated the time at which the rock formed. So the geological time scale in table 10 represents interpretations from fossils and radioactivity. The earth is believed to be about 5.6 billion years old. The

TABLE 10

Geological Era	Beginning (years before present)	Duration (years)	Characteristic Life Forms
Cenozoic	70,000,000	70,000,000	mammals
Mesozoic	225,000,000	155,000,000	reptiles
Paleozoic	600,000,000	375,000,000	invertebrates
Precambrian	5,600,000,000	5,000,000,000	no life except algae

fossiliferous strata record only the last 11% of earth's history. And human civilization has lasted only 10,000 years, a brief moment on the geological time scale. The immensity of geological time is the major discovery of geology. There has been ample time for very slow processes to produce large consequences.

OCEANS cover 70% of the earth's surface. The salts in seawater were dissolved during the weathering of bedrock. The major ions in seawater are shown in table 11. During one period or another, every portion of the continents has been depressed beneath sea level, and most of the strata seen in roadcuts throughout the United States are marine deposits. A typical cross section for the Atlantic Ocean would look like figure 17. The broad, shallow continental shelf collects much sediment from the continents. The mid-ocean ridge has frequent earthquakes and volcanic eruptions, so it is thought to be the site of upwelling of material from within the mantle.

TABLE 11

Ion		Percent
Chloride	Cl^{-1}	55.1
Sodium	Na^{+1}	30.6
Sulfate	SO_4^{-2}	7.7
Magnesium	Mg^{+2}	3.7
Calcium	Ca^{+2}	1.2
Potassium	K^{+1}	1.1
Bicarbonate	HCO_3^{-1}	0.4

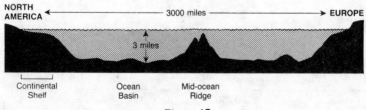

Figure 17

CONTINENTS may be divided into two zones. A shield area is a broad plain of Precambrian-age granite and gneiss, providing a stable core for the continent. Most of Canada is such a shield area. The second type of zone is one of recent upheaval, an orogenic (mountain-building) zone. Mountains occur in long, narrow belts, mostly along the edges of continents. On a map or globe, look at the western margins of South and North America to realize the edge-of-continent location of mountain ranges. Orogenic belts are sites of many earthquakes and volcanoes.

EARTH MOVEMENTS are the result of forces within the earth, where temperature and pressure differences lead to instability. The stress is particularly severe in orogenic zones, which are characterized by volcanism, metamorphism, deformation, and uplift. Two styles of rock deformation are faulting and folding as shown in figure 18. Today many geologists attribute edge-of-continent deformation to plate tectonics, a modern theory that suggests that oceanic crust emerges from the mantle along the oceanic ridges. Broad plates of oceanic crust may spread outward from the ridges until they crush against the margins of continental crust, the oceanic plate being forced under the continental plate back into the mantle.

STRUCTURES IN LAYERED ROCKS

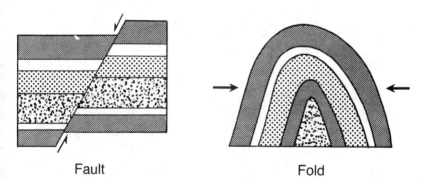

Fault　　　　　　　　　　Fold

arrows show movements
Figure 18

NATURAL RESOURCES obtained from the earth may be classed as metal and fuel deposits. Most metals are obtained by mining ore minerals in open-pit or underground mines. Gold is the only metal to occur commonly in its native, metallic state. Precambrian strata rich in chemically precipitated iron oxides are mined in northern Minnesota. Copper sulfides were deposited in veins by hydrothermal (hot water) solutions following igneous intrusions in Arizona, New Mexico, and Utah. By contrast, the fuel deposits (except uranium) are of organic origin. Plant debris that accumulated in ancient swamps has been transformed by heat and pressure to coal beds. Other plant and animal remains yielded oil and natural gas, which have accumulated in porous reservoir rocks. Uranium deposits are inorganic chemical precipitates from groundwater.

METEOROLOGY is the science of the atmosphere and weather. The composition of air is shown in table 12. The amount of water vapor in the air depends on the prevailing temperature and the availability of water. The hydrologic cycle as shown in figure 19 links the processes. Water in the ocean is evaporated by the sun's energy. As the warm, moist air rises to altitudes of lower pressure, the air expands and cools. The cooling results in condensation of water vapor to minute droplets of liquid water, which make up visible clouds. The clouds may be blown inland and cool further, leading to precipitation of rain or snow. Much of such precipitation is drained by rivers (the process is termed *runoff*) into the ocean, but an important fraction seeps into porous soil and rocks. That groundwater percolates laterally through permeable materials until it too ultimately reaches the ocean.

THE HYDROLOGIC CYCLE

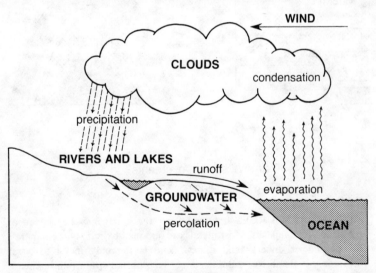

Figure 19

TABLE 12

Gas		Percent
Nitrogen	N_2	78.08
Oxygen	O_2	20.95
Argon	Ar	0.93
Carbon dioxide	CO_2	0.03
Water vapor	H_2O	varies

GLOSSARY OF TERMS IN GEOLOGY AND METEOROLOGY

ALLUVIUM: Loose sediment deposited by a stream.
ALTIMETER: An instrument for measuring altitude, commonly by means of air pressure.
ANEMOMETER: An instrument for measuring wind speed.
ANTICLINE: The crest of a rock fold.
AQUIFER: A bed of rock permeable to underground water.
ARTESIAN WELL: A water well drilled into a confined aquifer with high water pressure which forces the water toward the surface.
ATMOSPHERIC PRESSURE: The pressure exerted by the weight of the air lying directly above the area; at sea level, about 15 pounds per square inch.
ATOLL: A coral reef enclosing a lagoon.
AURORA: The radiant emission from the upper atmosphere known as the northern lights in our continent.
BAROMETER: An instrument for measuring atmospheric pressure.
BASALT: An igneous rock formed from lava.
BATHOLITH: A large intrusion of igneous rock, more than 40 square miles in area.
CALDERA: A large volcanic crater from collapse of the summit.
CAVERN: A large cave usually formed by the dissolving action of groundwater.
CIRQUE: A spoon-shaped hollow eroded by a glacier at the head of a mountain valley.
CIRRUS: A high-altitude cloud type composed of delicate white streamers.
CLOUD: A collection of tiny water or ice droplets sufficiently numerous to be seen. See cirrus, cumulonimbus, cumulus, nimbus, and stratus.
COAL: A rock composed of partly decayed and compressed plant material.

CONGLOMERATE: A sedimentary rock consisting of pebbles cemented together.
CONTINENTAL DRIFT: The hypothesis of continents moving laterally.
CONTINENTAL SHELF: The marine zone between the shore and the deep ocean basin.
CONTOUR: On a topographic map, a line connecting points of equal elevation. Hills are shown as concentric circles.
CORE: The center of the earth.
CRUST: The thin outer zone of the earth, above the mantle.
CUMULONIMBUS: A dark, puffy cloud type found in thunderstorms.
CUMULUS: A white, puffy cloud type with cauliflower-shaped tops.
CYCLONE: A low-pressure area around which winds blow counterclockwise in the Northern Hemisphere.
DEFORMATION: The tilting, bending, and breaking of rock layers.
DELTA: A triangular deposit of sediment at the mouth of a river.
DIASTROPHISM: A general term for large-scale earth movements.
DIKE: A tabular body of intrusive rock, cutting across older rock.
DIP: The angle of tilt of a rock layer, measured in degrees.
DIVIDE: The ridge between areas draining into different river systems.
EPICENTER: The place on the earth's surface closest to the origin of an earthquake.
EROSION: The removal of rock debris by water, ice, and wind.
EVAPORATION: The process by which a liquid changes to a gas; specifically, when water changes to water vapor.
EXTRUSIVE: Igneous rock of volcanic origin.
FAULT: A planar break in rock along which displacement has occurred.
FIORD: A glacial valley drowned by the sea.
FOLD: Bent or warped rock layers.
FORMATION: A rock unit shown on a geological map.
FOSSIL: Preserved trace of ancient life.
FRONT: The boundary between two air masses of different temperatures; a common site for cloud formation and precipitation.
GABBRO: A coarse-grained igneous rock, dark in color.
GEOSYNCLINE: The zone where a mountain chain forms.
GEOTHERMAL ENERGY: Heat obtained from hot water or steam within the earth.
GEYSER: A hot spring which periodically erupts steam and hot water.
GLACIER: A body of permanent ice thick enough to slowly flow under its own weight.
GNEISS: A coarse metamorphic rock.
GRANITE: A coarse-grained igneous rock, light in color from its quartz and feldspar grains.

GROUNDWATER: Subsurface water in the pores and cracks of soil and rocks.
HARDNESS: The relative resistance of a mineral to scratching, usually expressed on Mohs' scale of 1 (softest) to 10 (hardest).
HUMIDITY: The amount of water vapor in the air.
HUMUS: The organic matter in the soil.
HURRICANE: A large, severe tropical storm having wind speeds exceeding 73 miles per hour.
HYGROMETER: An instrument for measuring humidity.
IGNEOUS: Rock formed by the solidification of molten rock material.
INTRUSIVE: Igneous rock crystallized beneath the surface of the earth.
ISOBAR: On a weather map, a line of equal pressure.
ISOTHERM: On a weather map, a line of equal temperature.
LATITUDE: Location north or south of the equator, expressed in degrees.
LAVA: Molten rock at the surface of the earth, near a volcanic vent.
LIMESTONE: Sedimentary rock composed of calcium carbonate.
LITHIFICATION: The compaction and cementation of loose sediment.
LONGITUDE: Location east or west of the prime meridian at Greenwich, England.
MAGMA: Molten rock within the earth.
MANTLE: The zone of the earth between the core and the crust.
MARBLE: A metamorphic rock derived from limestone.
MEANDER: A wide curve or bend in the course of a large, winding river.
METAMORPHIC: Rock formed by the transformation, under high temperature and pressure, of older sedimentary or igneous rock.
METAMORPHISM: The recrystallization of rock to new minerals or texture.
METEOROLOGY: The science of the atmosphere and weather.
MINERAL: A naturally occurring inorganic chemical compound.
MIRAGE: An optical illusion where the sky is seen in place of the land, which appears wet.
MOHO: The Mohorovičić discontinuity, the boundary between the mantle and the crust.
MORAINE: A ridge of rocks and mud deposited by a glacier.
NIMBUS: A term for clouds which are raining or snowing.
OOZE: Very fine sediment on the ocean floor.
ORE: A rock rich enough in an economically valuable mineral to be mined.
OROGENY: The process of mountain building.
PALEONTOLOGY: The science of fossil life.
PERMAFROST: Ground that is frozen throughout the year.
PETROLEUM: A liquid fuel from the transformation of plant and animal remains.

PILLOW LAVA: A basaltic lava flow resembling a pile of pillows as it was split by contact with seawater.
PLACER: A mineral deposit in the gravel of a streambed.
PLASTIC: Deformation by flowing of solid rock or ice under great pressure.
PLUTONIC: Igneous rock that has crystallized beneath the earth's surface, as opposed to *volcanic* rock.
PRECIPITATION: Any form of water, whether liquid or solid particles, that falls from the atmosphere; rain, sleet, snow, or hail.
RADIOMETRIC DATING: Determining geological age by measuring the amount of radioactive decay products in a rock or fossil.
RAINBOW: A circular arc of colored bands produced by the refraction and reflection of sunlight by a sheet of raindrops. The sun must be behind the observer.
RELIEF: The variation in elevation in an area.
RICHTER SCALE: A scale measuring earthquake magnitude.
SANDSTONE: A sedimentary rock consisting of sand grains cemented together.
SCHIST: A medium-grained metamorphic rock.
SEAMOUNT: An underwater volcanic peak, commonly with a flat top.
SEDIMENTARY: Rock formed by deposition at the earth's surface.
SEISMIC: Refers to earthquakes.
SEISMOGRAPH: An instrument used to detect movements of the earth's crust.
SHALE: A sedimentary rock consisting of mud-sized particles.
SILL: A tabular body of igneous rock, parallel to sedimentary strata.
SILT: Sedimentary particles between clay and sand in size.
SLATE: A fine-grained metamorphic rock that splits into smooth sheets.
SOIL: Broken and decomposed rock and humus.
SPECIFIC GRAVITY: Relative density; the density of a substance divided by the density of water, which therefore has a specific gravity of 1.
STALACTITE: A cone of calcareous rock hanging from the roof of a cavern.
STALAGMITE: A pillar of calcareous rock rising from the floor of a cavern.
STRATA: Layers of sedimentary rock; singular is *stratum*.
STRATOSPHERE: The atmospheric shell above the troposphere; the stratosphere extends from 6 to 30 miles above the earth's surface.
STRATUS: A cloud type which forms a uniform, gray layer.
SYNCLINE: The trough of a rock fold.
TALUS: The rock debris at the base of a cliff.
TECTONIC: Refers to movements of the earth's crust.
THRUST: Horizontal movement of the crust.

THUNDERSTORM: A small storm with cumulonimbus clouds accompanied by lightning and thunder, rain, and gusty winds.

TILL: Rocks and mud deposited by a glacier.

TOPOGRAPHY: The shape of the earth's surface.

TORNADO: A violently rotating column of air extending as a funnel beneath a cumulonimbus cloud. The winds may attain a speed of 200 miles an hour.

TROPOSPHERE: The lowest 6 miles of the atmosphere, characterized by temperature decreasing with height.

TSUNAMI: A giant wave, miscalled a *tidal wave,* caused by an earthquake.

UNCONFORMITY: An erosion surface between two formations; represents a time gap in the stratigraphic sequence.

UNIFORMITARIANISM: The principle that the present is the key to the past.

VOLCANIC: Rock formed when lava reaches the surface.

WATERSHED: The area which drains into one river system.

WATER TABLE: The upper limit of groundwater, below which all pores in the rocks are filled with water.

WEATHERING: The physical and chemical destruction of rock by the atmosphere.

WESTERLIES: The dominant west-to-east winds over middle latitudes.

WIND DIRECTION: A wind is described by its source; for example, a north wind comes from the north and blows southward.

CHEMISTRY REVIEW

BASIC CONCEPTS

CHEMISTRY is the science of the substances which make up our world. Any tangible matter is chemical in its nature. Chemists study substances by measuring their properties and observing the changes they undergo. Matter is anything which occupies space and has mass. Matter may exist in the state of solid, liquid, or gas, depending on the prevailing temperature and pressure. Modern chemistry began when the Frenchman Lavoisier stated (1780) the law of conservation of mass: there is no gain or loss of mass in a chemical change. For example, when hydrogen reacts with oxygen to yield water, the water produced has precisely the same weight as the gases that reacted. Consequently, water can be considered to be built from the *elements* hydrogen and oxygen. Elements are the simplest chemical substances. Compounds (like water) are homogeneous substances of constant composition that have formed by chemical union of elements.

ATOMS are the smallest particles of chemical elements. The experimental fact that chemical compounds have fixed compositions led the Englishman John Dalton to propose (1805) that substances consist of small particles of matter, which he called atoms from the Greek word for indivisible. Later it was realized that chemical compounds must likewise have a smallest particle, the molecule. Water, for example, is made up of many molecules with the composition H_2O and the structure (arrangement) shown in figure 20. The straight lines denote bonds between atoms. One water

WATER MOLECULE

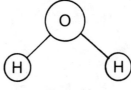

Figure 20

molecule is built from two atoms of hydrogen and one atom of oxygen. The molecules of the gaseous elements hydrogen and oxygen are shown in figure 21. Each of these molecules is itself diatomic, with two atoms bonded together. Finally, we can diagram the reaction that produces water molecules. (See figure 22.) The size of each box portrays the fact that the volume of gas is proportional to the number of molecules, rather than atoms. The chemical reaction is seen to be primarily a change in the bonding of atoms.

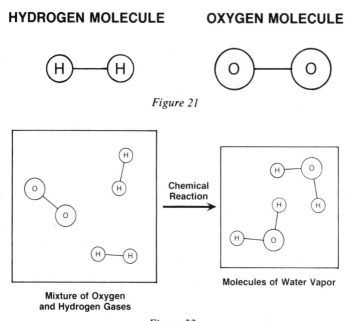

Figure 21

Figure 22

SUBATOMIC PARTICLES have been discovered as physicists learned how to split the supposedly indivisible atom. The three principal particles are listed in table 13. The masses are in international atomic mass units (amu). Protons and neutrons have masses of approximately 1 amu, while the electron is much lighter, zero for chemical purposes. An atom has a small, dense nucleus composed of protons and neutrons packed together. The nucleus is surrounded by a much larger cloud of electrons. Thus atomic weight is determined by the nucleus but the size of the atom is fixed by the electrons. *The number of protons establishes the chemical nature of an atom.* The number of neutrons is approximately equal to the number of protons, and the function of neutrons appears to be to stabilize the nucleus. The number of electrons precisely equals the number of protons for an electrically neutral atom.

TABLE 13

Particle	Charge	Mass
Proton	+1	1.007
Neutron	0	1.009
Electron	−1	0.001

TABLE 14

Element	Symbol	Protons	Neutrons	Mass
Hydrogen	H	1	0	1
Helium	He	2	2	4
Carbon	C	6	6	12
Oxygen	O	8	8	16
Iron	Fe	26	30	56
Uranium	U	92	146	238

CHEMICAL ELEMENTS may be assigned atomic numbers equal to the protons in their atoms. When elements are charted in order of increasing atomic number, a pattern of recurring physical and chemical properties is displayed. This periodic table of the chemical elements was first devised by the Russian Mendeleev (1869). Ninety-one elements occur naturally and another 12 have been made in the laboratory. Of these 103 elements, only the first 20 are shown in table 15. The elements increase in atomic number and atomic weight horizontally. Each vertical column contains a group of elements with similar chemical properties. Element behavior repeats every 8 atomic numbers, the octet rule. Lithium, sodium, and potassium are alkali metals that are very reactive and form caustic bases in solution. (Hydrogen has unique properties and is grouped with the alkali metals only because they have the same valence number, to be discussed later.) Beryllium, magnesium, and calcium are less reactive metals. Moving to the right, we leave the metallic elements and meet the nonmetals. Interconnected carbon atoms are the skeleton for organic molecules, while silicon atoms link into a similar framework within natural crystals. Nitrogen and phosphorous are slightly reactive nonmetals. Fluorine and chlorine are highly reactive gases that form strong acids in solution. Finally, the elements in the righthand column—helium, neon, and argon—are inert, "noble" gases that form no chemical compounds.

TABLE 15
ABRIDGED PERIODIC TABLE OF THE CHEMICAL ELEMENTS

1 hydrogen H							2 helium He
3 lithium Li	4 beryllium Be	5 boron B	6 carbon C	7 nitrogen N	8 oxygen O	9 fluorine F	10 neon Ne
11 sodium Na	12 magnesium Mg	13 aluminum Al	14 silicon Si	15 phosphorous P	16 sulfur S	17 chlorine Cl	18 argon Ar
19 potassium K	20 calcium Ca						
+1	+2	+3	+4, −4	+5, −3	−2	−1	0

valence for each column

COMPOUNDS are written as formulas with standard symbols for the chemical elements. The subscript following each symbol shows the number of atoms of that type in one molecule or formula unit of the compound; absence of a subscript implies one atom. Therefore a molecule of the gas ethylene (C_2H_4) has two atoms of carbon and four of hydrogen. The common mineral calcite ($CaCO_3$) has one calcium, one carbon, and three oxygen atoms per formula unit. Most compounds contain metallic and nonmetallic elements in proportions so that their valences sum to zero. Check again the abridged periodic table and note that hydrogen has a valence of $+1$ while oxygen is -2; consequently they combine in the proportion 2:1 as the formula H_2O denotes. Sodium ($+1$) combines with chlorine (-1) to yield NaCl, common table salt. The valence number summarizes the chemical behavior of each element. Metals have positive valence while nonmetals are negative. Note that in a formula the metallic symbol is written first. Now you should try to write the chemical formula for magnesium fluoride; the answer is printed as a footnote.

BONDS between atoms are electronic in origin. Electrically neutral atoms may share electrons to form a covalent bond, as in a hydrogen molecule. (See figure 23.) The electrons are in shells around the nuclear protons. Atoms which have an electrical charge are called *ions*. A positive ion has lost electrons (which have a charge of -1, remember) and a negative ion has gained electrons. Ions of different charge have a strong electrostatic attraction or ionic bond.

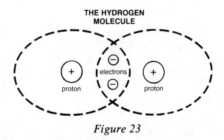

Figure 23

SOLIDS are characterized by their ability to retain their shape. They are relatively incompressible. Solids melt when heated and vaporize only slightly. All substances become solid if cooled sufficiently. Solids may be either crystalline or amorphous, depending on whether the arrangement of the atoms is regular or irregular. Figure 24 shows the crystalline structure of sodium chloride. Note that the positive sodium ions and the negative

---

Magnesium fluoride is MgF_2.

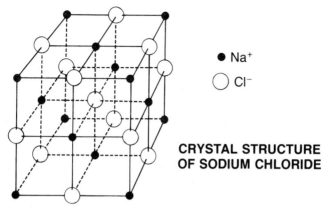

Figure 24

chlorine ions alternate in both the rows and columns, so that differently charged ions are closest neighbors. Ionic bonding holds the atoms in this rigid, stable arrangement.

LIQUIDS take on the shapes of their containers, yet cannot be compressed to any significant extent. The volume of a liquid is constant unless evaporation is occurring. Liquids crystallize when chilled sufficiently, while heat causes liquids to vaporize; boiling is very rapid vaporization. When evaporation occurs, some molecules in the liquid have gained enough energy to overcome the attractive forces exerted by the neighboring molecules and escape from the surface of the liquid to become gas. Thus the liquid state is intermediate between the solid and gaseous states, with regard to molecular motion and attractive forces between molecules.

GASES expand to fill any available space. A gas is a compressible fluid, with its volume determined by the pressure and temperature of the environment. The volume varies inversely with pressure, a relationship known as Boyle's law (see figure 25) and written

$$PV = k_1$$

where k_1 is a constant. Another relationship is Charles' law (see figure 25), that gas volume varies directly with temperature:

$$V = k_2 T$$

where k_2 is another constant, not equal to k_1. According to the kinetic molecular theory, gases are swarms of tiny molecules moving at a speed dependent on the temperature. Pressure is due to the molecular impacts on the walls of the container. The Italian physicist Avogadro proposed that equal numbers of molecules are contained in equal volumes of all gases, providing the pressure and temperature are identical. Experiments have found that 22.4 liters of gas at room temperature and pressure contain 6×10^{23} molecules, a value known as Avogadro's number.

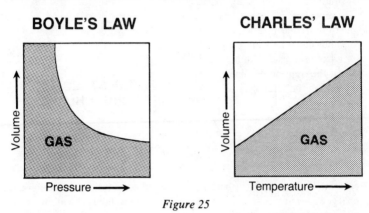

Figure 25

WATER is the most familiar of all liquids. It is a major constituent of living creatures and our environment. Ordinary water, even rainwater, is impure, with dissolved salts and gases. It may be purified by distillation, where the water is boiled to vapor, which is condensed and collected. The particular *state* of water—liquid, solid, or gas—is determined by the pressure and temperature. (See figure 26.) The dashed line shows the behavior of water at room pressure (1 atmosphere), freezing at 0°C and boiling at 100°C. At other pressures, the freezing and boiling temperatures for liquid water differ from the familiar values.

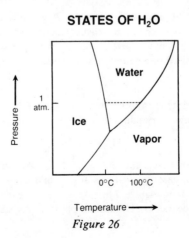

Figure 26

A SOLUTION is a mixture of two or more substances, the proportions variable between wide limits. Most solutions have a liquid solvent containing a lesser amount of dissolved solute, either a solid, a gas, or another liquid. An aqueous solution has water as the solvent. The concentration of a solution expresses the amount of solute dissolved in a standard unit (usually 1 liter) of the solvent. The solubility of a substance is the maximum concentration of solute which a solution can hold. The solubility of solids commonly increases with higher temperature, while the solubility of gases decreases with higher temperature. Precipitation of a salt occurs when the solution becomes supersaturated in that salt, the concentration exceeding the solubility.

IONS form when compounds dissociate in solution to positive and negative particles. For example, magnesium fluoride (MgF_2) separates in solution to magnesium and fluorine ions:

$$\underset{salt}{MgF_2} \xrightarrow{H_2O} \underset{ions\ in\ solution}{Mg^{+2} + 2F^{-1}}$$

In many cases, at least one ion is complex, containing several atoms. An example where two complex ions are present is the dissociation of ammonium nitrate:

$$NH_4NO_3 \xrightarrow{H_2O} NH_4^{+1} + NO_3^{-1}$$

The strong, covalent nitrogen-hydrogen and nitrogen-oxygen bonds were not disrupted by the water. However, the solution did break the ionic bond between the ammonium group of atoms and the nitrate group. The most important complex ion is the hydroxyl ion (OH^{-1}) which may be formed by the dissociation of water molecules:

$$H_2O \rightarrow H^{+1} + OH^{-1}$$

ACIDS AND BASES are solutions with unusual concentrations of either hydrogen or hydroxyl ions, such a solution being acidic or alkaline (basic). The familiar acids contain hydrogen which is loosely bound, while the bases are hydroxides of alkaline metals. (See table 16.) Acids and bases

TABLE 16

Acids	Hydrochloric	HCl
	Nitric	HNO_3
	Sulfuric	H_2SO_4
Bases	Sodium Hydroxide	NaOH
	Potassium Hydroxide	KOH
	Ammonium Hydroxide	NH_4OH

dissociate in aqueous solution and modify the concentration of hydrogen ions, which is measured as pH by litmus paper or other indicators. (See table 17.) The reaction of an acid with a base is called neutralization and releases some heat.

TABLE 17

Solution	pH
strong acid	1
weak acid	4
neutral	7
weak base	10
strong base	13

ELECTROLYSIS occurs as an ionic melt or solution conducts electricity. Pure water is a fine insulator, but may be made a conductor by adding an electrolyte, a substance that dissociates to many ions capable of transporting electric charge. Figure 27 shows the electrolysis of a dilute solution of sodium chloride. The positive sodium ions (cations) move toward the negative electrode (cathode), while the negative chlorine ions (anions) are attracted to the positive electrode (anode). At the electrodes the electrical charges react with water to yield oxygen and hydrogen gases, the volume of hydrogen being twice that of oxygen. So the dissolved salt enables a current to decompose water. Very pure oxygen and hydrogen may be

ELECTROLYSIS

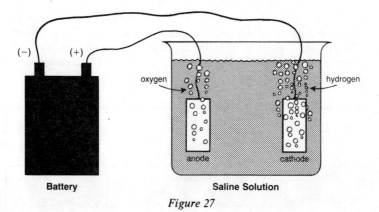

Figure 27

obtained by electrolysis of aqueous solutions. Other chemical elements may be obtained by passing a current through a molten salt. Electrolysis of fused sodium chloride (of course, no water is present) produces a layer of sodium metal on the cathode and bubbles of chlorine gas at the anode.

CHEMICAL REACTIONS show the number of molecules or formula units of the reactants and products. For example, nitrous oxide is a colorless, odorless gas which causes mild hysteria when breathed, hence the name *laughing gas;* it is prepared by heating ammonium nitrate crystals:

$$\underset{\underset{\text{ammonium nitrate}}{\text{Reactant}}}{NH_4NO_3} \xrightarrow{\text{heat}} \underset{\underset{\text{water vapor}}{\quad}}{2H_2O} + \underset{\underset{\text{nitrous oxide}}{\quad}}{N_2O}$$
(Products)

Conditions necessary for the reaction to proceed—like heat, light, or catalyst—are shown alongside the reaction arrow, which always points from reactants to products. In our example, one molecule of ammonium nitrate yielded two molecules of water and one of nitrous oxide. Now let's consider the atomic weights involved as shown in table 18. Adding up the atoms in the molecules, you find that the reaction states that, say, 80 grams of NH_4NO_3 yields 36 grams of H_2O and 44 grams of N_2O. Here we've returned to the conservation of mass, the great discovery of Lavoisier.

TABLE 18

Chemical Element		Atomic Number	Atomic Weight
Hydrogen	H	1	1
Nitrogen	N	7	14
Oxygen	O	8	16

GLOSSARY OF TERMS IN CHEMISTRY

ACID: A compound which yields hydrogen ions in solution.
ALCOHOL: An organic compound with the −COH functional group; the simplest alcohol is methanol, CH_3OH.
ALKALI METALS: Lithium, sodium, and potassium; common as hydroxides.
ALKALINE EARTH METALS: Beryllium, magnesium, and calcium; common as oxides.

ALLOY: A metal composed of two or more metallic elements.

AMORPHOUS: Describes the irregular structure of a noncrystalline solid.

ANION: An ion with a negative charge; in electrolysis, an anion moves toward the anode.

ANODE: The positive electrode.

ATOM: The smallest particle of matter that cannot be subdivided by chemical reactions.

ATOMIC NUMBER: The number of protons in an atomic nucleus; the different chemical elements have different atomic numbers.

ATOMIC WEIGHT: The weight of an atom relative to a carbon atom; $C^{12} = 12$. The atomic weight is approximately the number of protons plus neutrons.

AVOGADRO'S NUMBER: 6×10^{23}, the number of molecules in a gram molecular weight of a substance.

BASE: A compound which yields hydroxyl (OH^-) ions in solution.

BOILING POINT: The temperature at which a liquid changes to a gas.

BOILING POINT ELEVATION: An increase in the boiling point of a solvent, proportional to the amount of solute.

BOYLE'S LAW: The volume of a gas varies inversely with pressure.

CARBON DIOXIDE: CO_2 is a colorless, noncombustible gas under normal conditions.

CATALYST: A substance which accelerates a chemical reaction, without itself being a reactant.

CATHODE: The negative electrode.

CATION: An ion with positive charge; in electrolysis, a cation moves toward the cathode.

CHARLES' LAW: The volume of a gas varies directly with temperature.

COLLOID: A suspension of tiny particles in a fluid.

COMBUSTION: Rapid oxidation that releases heat and light.

COMPOUND: A substance formed by the chemical union of several chemical elements.

CONCENTRATION: The amount of dissolved material in a given amount of solution.

CONDENSATION: The liquefaction of a vapor.

CONSERVATION OF MASS: The law that there is no gain or loss of mass in a chemical reaction.

COVALENT BOND: The bonding of two atomic nuclei by sharing several electrons.

CRYSTAL: A solid in which the atoms have a regular, geometric arrangement.

DECOMPOSITION: A chemical reaction in which a compound is broken down into simpler compounds or elements.

DENSITY: Mass per unit volume of a substance.
DEUTERIUM: An isotope of hydrogen with a neutron in the nucleus as well as the necessary proton; *heavy water* is D_2O.
DIFFUSION: The mixing of different substances, commonly in a liquid or gas.
DISSOCIATION: Separation of a compound in solution to positive and negative ions.
DISTILLATION: The process of purification in which an impure substance is heated to vapors, which are collected and condensed.
ELECTRODE: A charged pole of metal dipped into a liquid that conducts electricity.
ELECTRON: A negatively charged, subatomic particle; electrons form a cloud around the atomic nucleus. Electron movement constitutes electrical current.
ELECTROLYSIS: A chemical change brought about by an electric current; used to separate chemical elements.
ELECTROLYTE: A liquid that conducts electricity.
ELEMENT: A substance which cannot be decomposed to simpler substances.
EQUILIBRIUM: The point at which two opposing chemical reactions balance.
EVAPORATION: A change of state from a solid or liquid to a gas.
FREEZING POINT: The temperature at which a liquid changes to a solid.
FREEZING POINT DEPRESSION: The decrease in freezing point of a solvent, proportional to the amount of solute.
FUSION: The melting of a solid to a liquid.
GRAPHITE: Soft, flaky crystalline form of carbon.
GROUP: A vertical column of elements in the periodic table; elements in one group have similar chemical properties.
HALOGENS: Fluorine, chlorine, bromine, and iodine; highly reactive nonmetals.
HARD WATER: Water containing dissolved salts of calcium or magnesium.
HETEROGENEOUS: Describes a substance that is a mixture.
HYDROCARBONS: Compounds of carbon and hydrogen.
HYDROLYSIS: Chemical decomposition of a compound by reaction with water.
ION: A charged atom or group of atoms formed by the gain or loss of electrons.
IONIC BOND: The electrical force of attraction between a positive ion and a negative ion, commonly within a crystal.
IONIZATION: Adding or subtracting electrons from an atom.

ISOTOPE: Isotopes of an element have the same number of protons and show the same chemical behavior, but they differ in number of nuclear neutrons and thus in atomic weight; isotopes may be stable or radioactive.
LITMUS: Paper that turns red in acid and blue in alkaline solution.
MELTING POINT: The temperature at which a solid changes to a liquid.
METALS: Elements that tend to lose electrons and form cations.
METHANE: The simplest hydrocarbon, a gas with the composition CH_4.
MISCIBLE: Capable of being mixed, referring to two liquids.
MIXTURE: Substances mixed without a chemical reaction; the substances can be in any proportion.
MOLE: Abbreviation for molecular weight.
MOLECULAR WEIGHT: The sum of the weights of the atoms in a compound.
MOLECULE: The smallest particle of a compound, composed of several bonded atoms.
NEUTRALIZATION: The chemical reaction of an acid and a base to form a salt and water.
NEUTRON: A subatomic particle of zero charge which occurs in the atomic nucleus.
NOBLE GASES: Also called inert gases, the nonreactive group of elements helium, neon, argon, and krypton.
NONMETALS: Elements which tend to gain electrons and form negative ions.
NUCLEUS: The dense center of an atom, made up of protons and neutrons.
OCTET RULE: The law that chemical properties tend to repeat in a period of eight atomic numbers.
ORGANIC COMPOUND: A compound with interconnected carbon atoms.
OXIDATION: The addition of oxygen to a substance.
OXIDE: A compound of oxygen and another element.
OZONE: A gas with molecules containing 3 oxygen atoms, O_3.
PERIODIC LAW: The chemical and physical properties of the elements are periodic functions of their atomic numbers.
pH: A number indicating the concentration of hydrogen ions in a solution. A pH of 7 is neutral, less than 7 is acidic, and greater than 7 is alkaline.
POLYMER: A giant organic molecule, made by chaining together smaller units.
PRECIPITATE: A solid which separates from solution.
PROTON: A subatomic particle with a positive charge, occurring in the atomic nucleus.
REACTION: A chemical change of substance, from reactant(s) to product(s).
REDUCTION: The removal of some oxygen from a compound.

REPLACEMENT: The release of an element from a compound by substituting a more active element.
SALT: A solid compound composed of both metallic and nonmetallic elements.
SATURATED: Describes a solution which contains as much solute as possible.
SOLUTE: The substance dissolved in a solution.
SOLVENT: The pure liquid within a solution.
STATE OF MATTER: Solid, liquid, or gas.
SUBLIMATION: The change from a solid to a gas, without an intermediate liquid.
SYNTHESIS: The formation of a compound by combining elements or simpler compounds.
TITRATION: Determination of the strength of a solution by adding a measured quantity of a known solution until the color changes or a salt precipitates.
TRANSITION ELEMENTS: The metallic elements in the middle of the long periodic table, including iron, manganese, copper, silver, and many other metals.
VALENCE: A number describing the combining power of an atom, the number of electrons it can gain or lose in combination with other atoms.
VAPOR: Gas.
VISCOSITY: The resistance of a liquid to flowage.
VOLATILITY: The ease of vaporization for a liquid or solid.

PHYSICS REVIEW

BASIC CONCEPTS

PHYSICS is the most basic and most general of the natural sciences. It covers subjects from matter to energy in the most general way. Physicists try to provide orderly explanations for natural events by formulating laws broad enough to explain all particular situations. Such laws are often suggested by regularities in experiments, but the clear logic and advanced mathematics used to construct physical formulas make physics the supremely theoretical science. The scientific method requires observation, conjecture, calculation, prediction, and testing. Successive scientific revolutions have taught us that today's laws are not certain, only more accurate than yesterday's laws.

MEASUREMENT is the beginning of scientific wisdom. The physicist's first reaction to a new idea is: Can it be measured? Can I describe it with numbers? Numerical data can be manipulated with many powerful mathematical tools, from arithmetic and geometry to statistics and differential equations. Physical quantities range from subatomic smallness to astronomic hugeness, so the numbers are conveniently expressed in *scientific notation,* in which any number is written in the form

$$N \times 10^P$$

where N is a number between 1 and 10, and P is a power of 10. The population of Brazil is about 130,000,000 and that number could be written as

$$1.3 \times 10^8$$

You should also be aware of the three basic units of the metric system. (See table 19.) A unit 1000 times the basic unit has the prefix *kilo,* so a kilometer equals 1000 meters. The prefix *milli* (as in millimeter) denotes a unit 1/1000 the basic unit.

TABLE 19

Quantity	*Unit*	*Symbol*	*Approximation*
Length	meter	m	1.1 yard
Volume	liter	l	1.1 quart
Mass	gram	g	1/30 ounce

MOTION is described by stating an object's position, velocity, and acceleration. Velocity is the rate of change of position with time. For example, an automobile that is 100 miles further along a highway at 3:00 than at 1:00 has an average velocity during that interval of

$$v = \frac{\Delta d}{\Delta t} = \frac{100}{2} = 50 \text{ miles/hour}$$

where the Δd represents change of distance and Δt is the change of time. Acceleration is the rate of change of velocity with time. If the automobile in our example had an initial velocity of 40 mph and a final velocity of 60 mph, then its average acceleration would be

$$a = \frac{\Delta v}{\Delta t} = \frac{20}{2} = 10 \text{ miles/hour/hour}$$

NEWTON'S LAWS relate the motion of an object to the forces acting upon it. The *law of inertia* asserts that, in the absence of any force, a body at rest will continue at rest while another body moving in a straight line will continue to move in that direction with uniform speed. Any change of speed or direction must be due to a force. The *law of acceleration* states that a body acted on by a force will undergo acceleration proportional to the force:

$$f = m \cdot a$$

where m is the mass of the object, the quantity of matter for that object. Table 20 lists masses for a range of objects. Newton's *law of reaction* says that for every action, there is an equal and opposite reaction.

TABLE 20

Representative Object	*Mass in Grams*
Electron	10^{-27}
Atom	10^{-23}
Amoeba	10^{-5}
Ant	10^{-2}
Human	10^{5}
Whale	10^{8}
Earth	10^{28}
Sun	10^{33}

GRAVITATION is familiar to us through weight, which is directly proportional to mass:

$$w = m \cdot g$$

where g is the acceleration due to gravity. The mass of an object is constant throughout the universe, but its weight varies with the object's position. Because the moon has a weaker gravitational field than the earth, an astronaut weighing 180 pounds here would weigh only 30 pounds there. Gravitational forces exist between all pairs of tangible objects, with the force being directly proportional to the product of their masses and inversely proportional to the square of the distance separating them. An astronaut weighs less on the moon because its mass is less than that of the earth. A balloon rising from the surface decreases in weight as it ascends and the distance from the earth increases.

ENERGY is the ability to perform work, to move objects. That ability can take several forms. The energy possessed by a moving object is called *kinetic* energy. An object in an unstable position has *potential* energy, for the position could be converted into movement. Let's consider a baseball thrown vertically upward. (See figure 28.) Its speed decreases upward because the acceleration of gravity is acting downward. The rising ball

CONSERVATION OF ENERGY

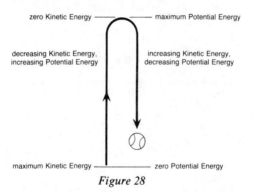

Figure 28

loses kinetic energy (slows down) as it gains potential energy (rises higher). At the peak of the ball's flight, the ball is instantaneously at rest, with no kinetic energy but maximum stored potential energy. As the ball falls, the potential energy is transformed into kinetic energy and the ball accelerates. *Thermal* energy also exists, for it has been shown that heat can be converted to motion, and motion can produce heat. Electricity and magnetism are still other forms of energy, for they can be converted into heat and motion. Notice that this key concept of energy is the abstract idea that there is something identical in motion, heat, and electricity, which appear so different to our senses. It is possible to define the various forms of energy so that their mathematical sum is constant. The law of conservation of energy states that energy can be neither created nor destroyed.

TEMPERATURE is a measure of the movement of the molecules in a substance. Heat is nothing else than kinetic energy on an atomic level. The basic temperature scale of science is the centigrade (or Celsius) scale (see table 21), on which water freezes at 0°C and boils at 100°C. At absolute zero, all molecular motion would cease. You should be able to convert our familiar Fahrenheit temperatures to centigrade or vice versa:

$$C = 5/9 \,(F - 32)$$
$$F = 9/5 \,C + 32$$

Just for practice, try to calculate the centigrade temperature equivalent to 50°F, then look at the answer in the footnote. The quantity of heat contained in a substance is measured in calories and must not be confused with temperature. At one temperature, a large mass of lead contains more heat than a smaller mass of lead. If those two pieces of lead had equal heat contents, the small mass would have a higher temperature than the larger mass.

TABLE 21

Temperature of Object in °C	
Absolute zero	−273
Oxygen freezes	−218
Oxygen liquifies	−183
Water freezes	0
Human body	37
Water boils	100
Wood fire	830
Iron melts	1535
Iron boils	3000

$C = 5/9 \,(F - 32) = 5/9 \,(50 - 32) = 5/9 \,(18) = 10°C$

SOUND is produced by the mechanical disturbance of a gas, liquid, or solid. The disturbance consists of alternating zones of abundant and scarce molecules, and such zones travel as waves. (See figure 29.) In such waves, the molecules vibrate and collide with each other, thus passing on their kinetic energy without changing their average position. The speed of sound depends on the physical properties of the medium through which it travels. In air, sound travels at 740 miles per hour, while steel transmits sound 15 times faster. The intensity level of sound is commonly reported in decibels.

SOUND WAVES IN A PIPE

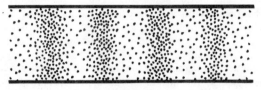

Figure 29

ELECTRICITY exists where the number of negative electrons does not precisely equal the number of positive protons. If we suspend small charged balls on nonconducting threads (see figure 30), we find that there are forces of repulsion between similar charges and attraction between

ELECTRIC FORCES

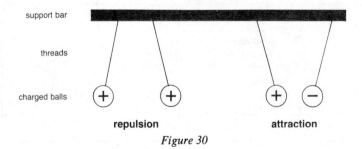

Figure 30

unlike charges. One type of electricity can neutralize the effect of the other type, so we regard the opposite charges as being due to an excess ($-$) or deficit ($+$) of *negative* electrons. Two bodies of opposite charge are said to have a difference of electrical potential, which is measured in volts. The chemical reaction in a battery produces a potential difference, and connecting the two poles through a conducting wire leads to the passage of an electric current, which is simply a flow of electrons.

MAGNETISM is displayed by permanent magnets and around electric currents. All of us have had the opportunity to study the interesting properties of permanent magnets, small bars or horseshoes of iron which have aligned internal structures induced by other magnets. The north pole of one magnet attracts the south pole of another, but like poles repel each other. Either pole can attract unmagnetized iron objects. Iron filings spread on a piece of paper above a bar magnet become arranged in a pattern which maps a magnetic *field* in the space around the magnet. (See figure 31.) The earth's magnetic field orients the iron needles of navigational compasses. An electric current also generates a magnetic field, demonstrating an intimate connection between electricity and magnetism. Later work has united these phenomena and light, too, into electromagnetic radiation.

**MAGNETIC FIELD
ABOVE A BAR MAGNET**

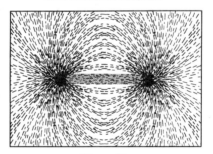

Figure 31

LIGHT seems to travel in perfectly straight lines as rays. The direction of a ray changes at the interface between two transparent materials, like air and water as depicted in figure 32. Notice that some of the light is

REFLECTION AND REFRACTION

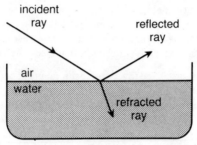

Figure 32

reflected, the angle of reflection being equal to the angle of incidence. The portion of the light that crosses the boundary is, however, deflected in another direction, and the angle of refraction does not equal the angle of incidence. Other optical experiments are inconsistent with a simple ray theory and require that light travels as waves of electromagnetic energy. When white light (including sunlight) is refracted by a glass prism, it is separated into its component colors as a beautiful spectrum. (See figure 33.) Experiments have shown that the various colors travel at the uniform speed c:

$$c = 186,000 \text{ miles / second} = 3 \times 10^8 \text{ meters / second}$$

DISPERSION OF SUNLIGHT

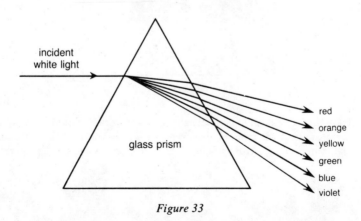

Figure 33

The colors differ in wavelength, and table 22 displays the relative wavelengths for all forms of electromagnetic energy.

TABLE 22
ELECTROMAGNETIC RADIATION

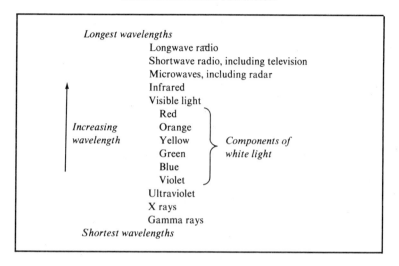

RELATIVITY of basic physical concepts like mass, distance, and time was postulated by Einstein (1905) to explain the experimental discovery that the speed of light in a vacuum was a universal constant. The principle of relativity states that all the laws of physics have the same form despite the movement of the observer. An observer performing an experiment on an object moving relative to himself finds that the object's measured mass is greater and length is less than if the object were at rest relative to himself. These curious effects are significant only at incredibly high velocities and may be neglected for most purposes. Einstein's famous law for the conversion of mass to energy

$$E = mc^2$$

suggested that atomic reactions could release unprecedented quantities of energy.

NUCLEAR ENERGY has been obtained by two different means, fission and fusion. Nuclear fission releases energy when a heavy nucleus splits into smaller fragments. (See figure 34. Black balls show neutrons, and white balls show protons.) Bombarding uranium with a neutron produces an unstable intermediate, which disintegrates to lighter nuclei with the conversion of 0.1% of the mass into energy. Nuclear fission is used in power plants and atomic bombs. The opposite process of nuclear fusion yields energy when very light nuclei unite to a heavier nucleus. (See figure 35.) A hydrogen bomb contains the two heavy isotopes of hydrogen, deuterium (H^2) and tritium (H^3) which unite to form helium nuclei and neutrons, with a conversion of 0.4% of the initial mass into energy. Stars (including the sun) derive their energy from nuclear fusion.

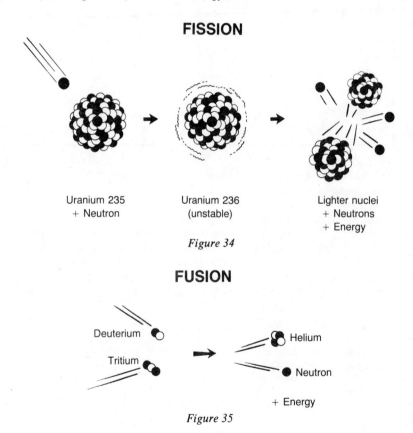

Figure 34

Figure 35

ASTRONOMY is the science of the heavens, from the moon to distant galaxies. In past cultures, men spent much more time outside than we do, and the recurring patterns of day and night, the seasons, eclipses, and the paths of the planets across the background of fixed stars led to the birth of science, an attempt to find order rather than mystery in our world. The ancient Greeks understood the cause of eclipses. A lunar eclipse darkens the moon as the earth passes between it and the sun, casting a shadow on the moon. (See figure 36.) A solar eclipse takes place when the moon passes between the earth and the sun, the moon blocking the sunlight for about 2 minutes for the most magnificent sight in astronomy. (See figure 37.) The first great triumph of modern science was when Copernicus (1543) realized that the earth was not at the center of the universe, but revolved around the sun yearly and rotated on its axis daily. Table 23 shows current data on the solar system. The earth is 93 million miles from the sun and has a radius of 4000 miles. From the table one can calculate that Jupiter, for example, is about 500 million miles from the sun and has a radius of about 44,000 miles; the mass of that giant planet is nearly three times that of all other planets combined.

LUNAR ECLIPSE

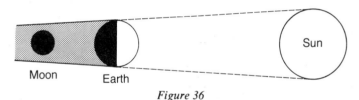

Figure 36

SOLAR ECLIPSE

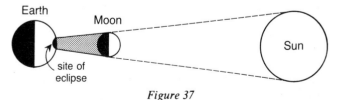

Figure 37

TABLE 23

Planet	Distance* from Sun	Radius*	Mass*	Moons
Mercury	0.39	0.38	0.05	0
Venus	0.72	0.96	0.82	0
Earth	1.00	1.00	1.00	1
Mars	1.52	0.53	0.11	2
Asteroids	2.8	(thousands of small bodies)		
Jupiter	5.2	11.0	317.8	15
Saturn	9.5	9.2	95.1	10
Uranus	19.2	3.7	14.5	5
Neptune	30.1	3.5	17.2	2
Pluto	39.5	0.5	0.1(?)	1

*relative to Earth = 1

GLOSSARY OF TERMS IN PHYSICS AND ASTRONOMY

ABSOLUTE TEMPERATURE: Temperature expressed in degrees Kelvin (°K), which is zero at absolute zero; °K = °C + 273.

ABSOLUTE ZERO: The lowest possible temperature, equal to −459°F, −273°C, or 0°K.

ACCELERATION: The rate of change of velocity with time.

ALPHA PARTICLE: A positive particle composed of two protons and two neutrons, emitted during radioactive decay.

AMPERE: The unit of measurement of electric current.

ARCHIMEDES' PRINCIPLE: A body immersed in a fluid is lifted by a force equal to the weight of the fluid displaced by the body.

ASTEROID: A planetary fragment or minor planet. Most of the thousands of known asteroids are between the orbits of Mars and Jupiter. Ceres is the largest asteroid.

AXIS: A straight line around which a body rotates.

BETA PARTICLE: An electron emitted from an atomic nucleus during radioactive disintegration.

BUOYANCY: The upward force on an object immersed in a fluid.

CALORIE: The amount of heat required to raise the temperature of one gram of water 1°C.

CAPILLARITY: The ability of liquids to rise in very thin tubes.

CENTRIFUGAL FORCE: The apparent force deflecting a rotating mass radially outward.

CHAIN REACTION: Occurs when the fission of one atom causes the fission of other atoms.
CHARGE: The electrical state of matter, positive or negative.
CLOUD CHAMBER: A tank filled with vapor through which atomic and subatomic particles leave visible trails.
COMET: A diffuse body which glows with a prominent tail when its orbit brings it near the sun.
CONDUCTION: Transfer of heat or electricity.
CONSERVATION OF ENERGY: Energy may be changed from one form to another, but it can't be created or destroyed.
CONSTELLATION: An apparent group of stars.
CORONA: A halo of glowing gases around the sun, visible only during a total solar eclipse.
CRITICAL MASS: The amount of fissionable material necessary to sustain a chain reaction.
CURRENT: A direct electric current (DC) flows in one direction, but an alternating current (AC) periodically reverses the direction of flow.
CYCLOTRON: A machine that accelerates atomic particles by means of an alternating electric field.
ECLIPTIC: The plane within which the planets orbit around the sun.
ELECTROMAGNETIC RADIATION: Energy traveling as a disturbance of electric and magnetic fields.
ENERGY: The ability to perform work. Kinetic energy is due to a body's motion, while potential energy is due to a body's position.
ESCAPE VELOCITY: The speed a rocket must reach in order to leave the earth's vicinity.
FIELD: The region where a force is felt.
FISSION: The splitting of an atomic nucleus into parts.
FLUID: A liquid or gas.
FOCUS: The place where an image is formed by a mirror or lens.
FORCE: A force causes a body to accelerate.
FREQUENCY: The number of vibrations of a wave per unit time.
FULCRUM: The support on which a lever pivots.
FUSION: Nuclear fusion is the union of atomic nuclei.
GALAXY: An astronomical system composed of billions of stars; the Milky Way is our galaxy.
GAMMA RAY: High-energy electromagnetic radiation emitted during radioactive disintegration.
GIBBOUS: The phase of the moon which is more than half full.
GRAVITATION: The attraction of bodies because of their masses; an especially familiar case is the attraction between the earth and an object.
GROUND: A connection to the earth to dissipate an electric charge.
HALF-LIFE: The time required for the radioactivity of a substance to drop by one-half.

HEAT: Kinetic energy of molecular motion.
HYPOTHESIS: A tentative explanation of a phenomenon.
INERTIA: The ability of a body to resist acceleration, continuing either at rest or in motion with uniform velocity.
JUPITER: The fifth planet from the sun is the largest in the solar system, with a diameter eleven times that of the earth. Jupiter has fifteen known satellites, the largest being Ganymede.
LAW: A general statement about a group of related facts.
LENS: A transparent material shaped either concave or convex to refract light and form a magnified image.
LIGHT-YEAR: The distance light travels in one year.
MAGNETIC POLES: The ends of a bar magnet are referred to as the north and south poles.
MAGNITUDE: The apparent brightness of a star; a larger magnitude represents a dimmer star.
MARS: The fourth planet from the sun, with a diameter 53% that of the earth. Mars has two small satellites, Phobos and Deimos.
MASS: The quantity of matter; the measure of inertia.
MERCURY: The small planet closest to the sun, with a diameter 39% that of the earth.
METEOR: The streak of light produced by an interplanetary particle in its passage through the earth's atmosphere.
METEORITE: A rock from interplanetary space, found on the earth's surface.
MOMENTUM: The product of mass and velocity. The conservation of momentum is a fundamental law of nature.
NEBULA: A cloud of gas or dust in interstellar space.
NEPTUNE: The eighth planet from the sun, with a diameter nearly four times that of the earth. Neptune has two satellites, the largest being Triton.
NOVA: A star which suddenly becomes many times brighter than usual.
NUCLEUS: The central part of an atom, containing almost all of its mass; plural is *nuclei*.
ORBIT: The path of one celestial body around another.
PARALLAX: The apparent shift in position of an object viewed from two different points; used for range-finding by triangulation.
PAYLOAD: The load of explosives, instruments, or passengers carried by a rocket.
PENDULUM: A mass hanging on the end of a string or other support, which is free to swing.
PHOTON: A quantum or particle of light energy.
PLUTO: The ninth planet from the sun, discovered in 1930; Pluto has one satellite, Charon.

POLARIS: The star at the celestial north pole.
POLARIZED LIGHT: Light waves that vibrate in one plane, rather than in all directions.
PRISM: A glass wedge used to disperse light.
QUANTUM: A very small unit of energy.
QUASAR: Quasi-stellar radio source.
RADIATION: Includes both electromagnetic waves and particles emitted during radioactive disintegration.
RADIOACTIVITY: Spontaneous decay of an atomic nucleus, with emission of alpha particles, beta particles, and gamma rays.
RADIO TELESCOPE: An electronic antenna that receives radio waves from outer space.
REFRACTION: The bending of a light ray or wavefront at the boundary between two substances.
REVOLUTION: The movement of a celestial body in its orbit. The earth takes $365\frac{1}{4}$ days to revolve around the sun.
ROTATION: The spinning of a body on its axis. One rotation of the earth requires 24 hours.
SATELLITE: A body orbiting around a planet.
SATURN: The sixth planet from the sun has very prominent rings and a diameter over nine times that of the earth. Saturn has ten known moons, the largest being Titan.
SCIENCE: Systematic and verifiable knowledge.
SPECTROSCOPE: An instrument that separates a beam of light into its component colors, usually for obtaining a chemical composition.
SPECTRUM: The visible spectrum is the band of colors seen when white light is dispersed. The electromagnetic spectrum is the total range of frequencies of electric and magnetic waves, including radio and light waves.
TEMPERATURE: Average kinetic energy of a group of molecules; determines the direction of heat flow.
TERMINAL VELOCITY: The final, constant speed of fall of a body in a fluid.
TIDE: The rising and falling of the ocean due to the gravitational attraction of the moon and sun.
URANUS: The seventh planet from the sun, with a diameter almost four times that of the earth. Uranus has five known satellites and a faint ring.
VAN ALLEN BELTS: Zones around the earth rich in electrically charged particles.
VELOCITY: The rate of change of position with time.
VENUS: The second planet from the sun is very nearly the same size as the earth and is cloaked by dense, hot clouds.
VOLT: Unit of measurement of electric potential.

WAVE: A disturbance that can travel.
WAVELENGTH: The distance from one wave crest to the next.
WEIGHT: A measure of the gravitational pull on an object.
WEIGHTLESSNESS: A condition that arises where accelerating forces precisely offset one another.
WORK: The product of force and distance.
X RAYS: Short electromagnetic waves that can penetrate solids.
ZENITH: The point directly overhead.

PART V: Practice-Review-Analyze-Practice

This section contains two full-length practice simulation GED Science Tests. The practice tests are followed by complete answers, explanations, and analysis techniques. AN IMPORTANT AID IS THE EXPLANATION THAT FOLLOWS EACH QUESTION IN THIS GUIDE. Even if you were able to determine the correct answer, *read the explanation.*

The format, levels of difficulty, question structure, and number of questions are similar to those on the actual GED Science Test. The actual GED is copyrighted and may not be duplicated, and these questions are not taken directly from the actual tests.

When taking these exams, try to simulate the test conditions by following the time allotments carefully. Remember the total testing time is 90 minutes.

NOTE—Before you take the first practice test, you may wish to reread the section on how to take the test.

PRACTICE TEST 1
WITH COMPLETE ANSWERS AND EXPLANATIONS

ANSWER SHEET FOR PRACTICE TEST 1
(Remove This Sheet and Use It to Mark Your Answers)

#	① ② ③ ④ ⑤	#	① ② ③ ④ ⑤	#	① ② ③ ④ ⑤
1	① ② ③ ④ ⑤	26	① ② ③ ④ ⑤	51	① ② ③ ④ ⑤
2	① ② ③ ④ ⑤	27	① ② ③ ④ ⑤	52	① ② ③ ④ ⑤
3	① ② ③ ④ ⑤	28	① ② ③ ④ ⑤	53	① ② ③ ④ ⑤
4	① ② ③ ④ ⑤	29	① ② ③ ④ ⑤	54	① ② ③ ④ ⑤
5	① ② ③ ④ ⑤	30	① ② ③ ④ ⑤	55	① ② ③ ④ ⑤
6	① ② ③ ④ ⑤	31	① ② ③ ④ ⑤	56	① ② ③ ④ ⑤
7	① ② ③ ④ ⑤	32	① ② ③ ④ ⑤	57	① ② ③ ④ ⑤
8	① ② ③ ④ ⑤	33	① ② ③ ④ ⑤	58	① ② ③ ④ ⑤
9	① ② ③ ④ ⑤	34	① ② ③ ④ ⑤	59	① ② ③ ④ ⑤
10	① ② ③ ④ ⑤	35	① ② ③ ④ ⑤	60	① ② ③ ④ ⑤
11	① ② ③ ④ ⑤	36	① ② ③ ④ ⑤		
12	① ② ③ ④ ⑤	37	① ② ③ ④ ⑤		
13	① ② ③ ④ ⑤	38	① ② ③ ④ ⑤		
14	① ② ③ ④ ⑤	39	① ② ③ ④ ⑤		
15	① ② ③ ④ ⑤	40	① ② ③ ④ ⑤		
16	① ② ③ ④ ⑤	41	① ② ③ ④ ⑤		
17	① ② ③ ④ ⑤	42	① ② ③ ④ ⑤		
18	① ② ③ ④ ⑤	43	① ② ③ ④ ⑤		
19	① ② ③ ④ ⑤	44	① ② ③ ④ ⑤		
20	① ② ③ ④ ⑤	45	① ② ③ ④ ⑤		
21	① ② ③ ④ ⑤	46	① ② ③ ④ ⑤		
22	① ② ③ ④ ⑤	47	① ② ③ ④ ⑤		
23	① ② ③ ④ ⑤	48	① ② ③ ④ ⑤		
24	① ② ③ ④ ⑤	49	① ② ③ ④ ⑤		
25	① ② ③ ④ ⑤	50	① ② ③ ④ ⑤		

CUT HERE

PRACTICE TEST 1

Time: 90 Minutes
60 Questions

DIRECTIONS

The Science Test consists of 60 multiple-choice questions in two parts. In the first part are 20 questions based on background knowledge of scientific concepts. The second part contains 40 questions based on several brief reading passages. You have 90 minutes to complete both parts of this test. There is no penalty for guessing.

Part A: General Science Knowledge

The following questions are not based on a reading passage. Select the best answer based on your knowledge of the concepts of natural science.

1. The volume of a gas varies
   (1) with pressure but not temperature
   (2) directly with pressure and inversely with temperature
   (3) directly with temperature and inversely with pressure
   (4) directly with both pressure and temperature
   (5) inversely with both pressure and temperature

2. A protozoan never contains
   (1) chromosomes
   (2) cytoplasm
   (3) plastids
   (4) stamens
   (5) vacuoles

3. Fungi are *not* useful for
   (1) an antibiotic
   (2) brewing beer
   (3) flavoring pizza
   (4) making bread rise
   (5) pasteurizing milk

4. Avogadro's number is an estimate of the
   (1) absolute zero of temperature
   (2) atoms in a unit of matter
   (3) gravitational constant
   (4) strength of an acid or base
   (5) total stars in the universe

5. The direct function of enzymes is to
   (1) carry oxygen in the blood
   (2) control organs in the body
   (3) facilitate reproduction
   (4) promote chemical reactions
   (5) stimulate the nervous system

6. The bending of a houseplant toward a window is an instance of
   (1) a conditioned reflex
   (2) photosynthesis
   (3) refraction
   (4) taxonomy
   (5) tropism

7. The forces that produce the earth's tides are due to
   (1) centrifugal forces of the rotating earth
   (2) continental drift
   (3) daily changes in temperature
   (4) the earth's magnetic field
   (5) the moon's mass

8. Charles Darwin suggested that species evolve by means of
   (1) distribution
   (2) motivation
   (3) mutation
   (4) radiation
   (5) selection

9. The Mariner space probes discovered Martian
   (1) active volcanoes
   (2) canals
   (3) craters
   (4) life
   (5) moons

10. Hydrochloric acid and sulfuric acid both contain
    (1) chlorine
    (2) hydrogen
    (3) iron
    (4) oxygen
    (5) sulfur

11. Which of the following cities would have the lowest mean temperature during July?
    (1) Berlin
    (2) Buenos Aires
    (3) New Orleans
    (4) Rome
    (5) Toronto

12. A speed of 100 kilometers per hour is approximately equal to
    (1) 40 miles per hour
    (2) 60 miles per hour
    (3) 80 miles per hour
    (4) 90 miles per hour
    (5) 120 miles per hour

13. A dog and its fleas is one example of
    (1) embolism
    (2) metabolism
    (3) mitosis
    (4) osmosis
    (5) symbiosis

14. The sun's energy is chiefly caused by
    (1) chemical reaction
    (2) combustion
    (3) gravitation
    (4) nuclear fission
    (5) nuclear fusion

15. The chemical concept of valence concerns
    (1) atomic weights
    (2) electrolysis of solutions
    (3) the combining power of atoms
    (4) the vapor pressure of liquids
    (5) the volume of gases

16. For which of the following elements is the commercial supply obtained otherwise than by mining?
    (1) copper
    (2) iron
    (3) nickel
    (4) nitrogen
    (5) silver

17. All of the following are considered to be electromagnetic radiation *except*
    (1) electricity
    (2) microwaves
    (3) radio
    (4) sound
    (5) sunlight

18. Photosynthesis in plants does *not* require the availability of
    (1) carbon dioxide
    (2) chlorophyll
    (3) light
    (4) oxygen
    (5) water

19. Which of the following creatures is *not* an arthropod?
    (1) bee
    (2) crab
    (3) earthworm
    (4) shrimp
    (5) spider

20. pH is a measure of
    (1) the energy released by a chemical reaction
    (2) the pressure of a gas
    (3) the solubility of a salt
    (4) the strength of an acid
    (5) physical properties of metals

Part B: Science Reading Comprehension

This section contains 10 reading passages, with several questions about each passage. Read the passage first and then answer the questions following it. Refer to the passage as often as necessary in selecting the best answer. Remember, there is no penalty for guessing.

The chemical mechanisms of a food chain take the form of cycles. In the oxygen cycle, atmospheric oxygen is used by plants and animals in burning food (respiration). Burning food releases energy, water, and carbon dioxide. Carbon dioxide and water are used by green plants in photosynthesis, which releases oxygen, and the cycle starts anew. In the carbon cycle, which is linked with the oxygen cycle, photosynthesis converts carbon dioxide and water into food. Some of the food is used in respiration, which returns carbon dioxide to the atmosphere, and the remainder is forged into protoplasm. After the death of an organism, its protoplasm decays and releases carbon dioxide to the atmosphere. Carbon dioxide also enters the atmosphere through the combustion of organic fuels, like coal and oil. In the nitrogen cycle, microorganisms in the soil bring about the putrefaction of animal wastes and dead organisms. In this process ammonia is released, and another group of microorganisms (nitrifying bacteria) convert this ammonia into soluble nitrate salts. The roots of plants absorb these nitrates, and they are then used to manufacture proteins. Herbivorous animals convert these plant proteins into animal proteins. Denitrifying bacteria decompose nitrates into free nitrogen, a gas that escapes to the atmosphere, and nitrogen-fixing bacteria retrieve the lost nitrogen by converting atmospheric nitrogen into organic compounds useful to plants.

21. Photosynthesis does *not* involve which cycle(s)?
    (1) carbon only
    (2) nitrogen only
    (3) oxygen only
    (4) carbon and nitrogen
    (5) nitrogen and oxygen

22. Organic decay is part of which cycle(s)?
    (1) carbon only
    (2) nitrogen only
    (3) oxygen only
    (4) carbon and oxygen
    (5) all three cycles

23. Nitrogen gas is produced by
    (1) ammonia
    (2) decay
    (3) microorganisms
    (4) photosynthesis
    (5) respiration

24. Which of the following is apparently *not* essential for plant life?
    (1) carbon dioxide
    (2) light
    (3) water
    (4) nitrifying bacteria
    (5) denitrifying bacteria

An accelerator is a device which is used to speed a beam of nuclear particles toward a target. There are several types of accelerators. The linear accelerator is used to accelerate electrons, various ions, protons, deuterons, and alpha particles. The cyclotron accelerates particles in a magnetic field between two electrodes which change polarity at regular intervals, causing the particles to spiral outward in a circular path of increasing radius. It can accelerate protons, deuterons, and alpha particles. The synchrotron is used for electrons and protons, the betatron for electrons only. Neutrons cannot be accelerated because they have no charge. Slow neutrons, produced during artificial transmutations, have been used to initiate nuclear changes.

The bombardment of copper to form gallium

$$_{29}Cu^{63} + {}_2He^4 \rightarrow {}_{31}Ga^{66} + {}_0n^1$$

is typical of a class of reactions in which an atom is bombarded with an accelerated alpha particle to form another element with the release of a neutron.

25. What physical property of particles is used to accelerate them?
    (1) charge
    (2) mass
    (3) polarity
    (4) shape
    (5) size

26. Which of the following accelerators is not suited for protons?
    (1) betatron
    (2) cyclotron
    (3) linear accelerator
    (4) synchrotron
    (5) all four can accelerate protons

27. In the bombardment of copper (Cu) to produce gallium (Ga), what is the *alpha particle*?
    (1) copper nucleus
    (2) gallium nucleus
    (3) helium nucleus
    (4) neutron
    (5) proton

28. Each gallium nucleus produced must contain how many neutrons?
    (1) 0
    (2) 1
    (3) 31
    (4) 35
    (5) 66

Fossils are the remains or indications of ancient life on earth. They tell us what plants and animals have existed in the past, what climatic and geological environments they lived in, and the age of a fossil-bearing rock. Fossils vary from microscopic organisms to complete bodies of Ice-Age mammals. Age is not a criterion; skeletons of early man are regarded as fossils. Footprints are as much fossils as are bones. Ancient geological features that are not of organic origin are not fossils.

Fossils usually consist of the hard parts of organisms. They are often petrified, being filled or replaced by mineral matter deposited by percolating groundwater. Bones, teeth, scales, shells, and pollen are among the durable parts of animals and plants that are most likely to be preserved as fossils, while the softer parts decay and leave no trace.

The dead plant or animal must be buried quickly before it decomposes, is destroyed by erosion, or is eaten by scavengers. The environments of fossil preservation are many, although they differ in importance. Most fossils occur in marine sediment, where living forms were especially abundant. Some are found in freshwater sediment, deposited by streams and lakes. Bogs, tundra, volcanic ash, dunes, and caves provide opportunities for the burial and preservation of organisms.

29. Which of the following objects could *not* be a fossil?
    (1) cave stalactites
    (2) fish scales
    (3) prehistoric tool
    (4) shark skeleton
    (5) worm burrows

30. In the passage, the term *petrified* must mean
    (1) decayed
    (2) dissolved
    (3) frightened
    (4) preserved
    (5) squeezed

31. Most decay bacteria are aerobic, so plant remains are *least* likely to survive in
    (1) lake mud
    (2) peat bogs
    (3) sand dunes
    (4) swamp mud
    (5) volcanic ash

32. Which of these organisms is *least* likely to be preserved as a fossil?
    (1) fish
    (2) lobster
    (3) oyster
    (4) sea urchin
    (5) worm

Respiration includes the exchange of gases in the lungs between the blood and air, the exchange of gases between the blood and the tissues, and the transport of gases by the blood between the lungs and tissues. The gaseous exchange in the lungs is called external respiration, whereas the exchange in the tissues is referred to as internal respiration.

External respiration involves the inspiration (breathing in) and expiration (breathing out) of air. In the lungs, oxygen diffuses from the alveolar air into the blood, and carbon dioxide and water vapor diffuse from the blood into the alveolar air. The alveoli are the microscopic air sacs that line the lungs. Each gas diffuses because of pressure differences for that particular gas. For example, the partial pressure of oxygen in inspired air is about double that in venous blood, so that gas diffuses from the air in the alveoli into the blood.

Internal respiration in the tissues involves oxygen diffusing from the blood through the capillary walls into the intercellular fluid, while carbon dioxide passes in the reverse direction.

The transport of oxygen and carbon dioxide by the blood is almost entirely a matter of chemical combination rather than the simple solution of gases in the plasma. The blood carries oxygen chiefly as the unstable protein oxyhemoglobin, which is formed in the lungs and which decomposes into oxygen and hemoglobin in the tissues. Carbon dioxide is transported mostly as the compound sodium bicarbonate.

33. Which pair exchanges oxygen only indirectly?
    (1) air and lungs
    (2) air and tissues
    (3) blood and lungs
    (4) blood and tissues
    (5) they all exchange directly

34. Carbon dioxide is carried in the blood chiefly
    (1) as bicarbonate
    (2) as bubbles
    (3) by diffusion
    (4) in hemoglobin
    (5) in simple solution

35. Carbon dioxide passes from the veins to the alveoli because
    (1) hemoglobin is unstable
    (2) oxyhemoglobin is unstable
    (3) the alveoli are empty
    (4) the CO_2 pressure is lower in the alveoli
    (5) the CO_2 pressure is lower in the veins

36. Sodium bicarbonate may be inferred to form in the blood adjacent to the
    (1) alveoli
    (2) heart
    (3) lungs
    (4) thoracic cavity
    (5) tissues

Carbon is not very reactive at normal temperatures. At higher temperatures, it becomes quite reactive, combining with metals and nonmetals. For example, carbon burns in oxygen to form carbon monoxide (CO) or carbon dioxide (CO_2) once it is heated to its activation temperature, the product depending upon the amount of oxygen available.

Carbon monoxide results from the incomplete combustion of carbonaceous compounds. It is odorless and can kill by gradually interfering with the essential functioning of the blood. Carbon monoxide is a component of automobile exhaust.

Carbon dioxide is formed by the complete combustion of carbonaceous materials. It can be liquefied under pressure or solidified to form *dry ice*. The latter vaporizes at $-79°C$ without melting to a liquid. Carbon dioxide dissolves in water to form weak carbonic acid (H_2CO_3). Carbon dioxide and water vapor are the principal products of the burning of hydrocarbons like wood and oil. Carbon dioxide is also the product of organic fermentation:

$$\underset{sugar}{C_6H_{12}O_6} \xrightarrow{enzyme} \underset{ethanol}{2C_2H_5OH} + 2CO_2$$

The most common laboratory method of generating CO_2 is by the action of acid on carbonates or bicarbonates:

$$CaCO_3 + 2HCl \rightarrow CaCl_2 + H_2O + CO_2$$

37. *Complete combustion* apparently means complete
    (1) carbonization
    (2) fermentation
    (3) gasification
    (4) liquefaction
    (5) oxidation

38. A charcoal fire fanned by a brisk wind should yield mostly
    (1) $CaCl_2$
    (2) CO
    (3) CO_2
    (4) ethanol
    (5) water vapor

39. During fermentation, the enzyme apparently serves as
    (1) catalyst
    (2) acid
    (3) reactant
    (4) source of heat
    (5) sugar

40. The passage implies that dry ice has the composition
    (1) $CaCO_3$
    (2) CO
    (3) CO_2
    (4) C_2H_5OH
    (5) HCl

The geological activity of wind is usually associated with deserts, but dry regions need not be windy. A desert is an area of low precipitation, interior drainage, and a barren landscape. Deserts contain much more barren rock than they do deposits of sand.

Wind erosion is termed corrasion. It leaves basins called blowouts, and a pebbly surface known as lag gravel. The removal of the finer particles leaves coarser ones; the lag gravel may be termed a desert pavement. Beveled stones called ventifacts are shaped by wind abrasion. Ventifacts may be either glossy or pitted. Wind abrasion takes place close to the ground.

Wind is a more selective agent of transportation than moving water is, and much more so than moving ice. Eddies cause it to raise aloft some fine sediment, but most material is transported just above the surface. The lifting of fine material by wind is called deflation. The bouncing of sand along the ground is referred to as saltation. Particles roll and slide by traction.

Aeolian deposits are those made by wind. A reduction in energy and the presence of obstacles cause wind to deposit its load of sediment. Wind deposits include volcanic ash and sand dunes. Dunes migrate with the wind until they become fixed in position by the growth of vegetation.

41. Corrasion would *not* be inhibited by
    (1) barren rock
    (2) forest
    (3) high precipitation
    (4) lack of wind
    (5) low altitude

42. Particles undergoing traction are
    (1) beveled
    (2) fixed in place
    (3) larger than ventifacts
    (4) smaller than sand
    (5) too large for the wind to lift

43. Wind erosion would be very effective at
    (1) increasing temperature
    (2) killing vegetation
    (3) leveling dunes
    (4) moving gravel
    (5) separating dust, sand, and pebbles

44. A sand dune migrates mainly by means of
    (1) deflation
    (2) saltation
    (3) traction
    (4) vegetation
    (5) wave action

The retinal lining of the eye is made up of countless light-sensitive cells called rods and cones. The nerve fibers which emerge from these cells gather together at the optic disk and leave the eyeball as the optic nerve. The two optic nerves curve inward and meet just above the pituitary gland at the optic chiasma. Here there is a 50 percent "crossover" of nerve fibers; that is, the fibers from the inner half of the retina cross over to the optic nerve on the opposite side, while the fibers from the outer portion pass directly into the optic nerve on the same side. The optic nerves, upon reaching the cerebrum, diffuse into the visual areas of the two occipital lobes. Because of the crossover, each occipital lobe "sees" with both eyes.

The emission of nerve impulses by the retinal rods and cones results from photochemical reactions. Vision in dim light is caused by the photosensitivity of visual purple, a substance derived from the protein opsin and the vitamin A derivative retinine. Thus a deficiency of vitamin A causes a deficiency of visual purple which results in night blindness (nyctalopia).

Color vision involves three types of cones, each of which responds to one primary color—red, green, or blue. We experience red, green, or blue when the red, green, or blue cones (respectively) are stimulated by the corresponding frequencies of light. Other colors are seen as a result of the more or less intense stimulation of some combination of the cones. The sensation of white appears to result from the equal stimulation of all three types of cones.

45. The right-hand half of the brain receives information from
    (1) both eyes equally
    (2) chiefly the left eye
    (3) chiefly the right eye
    (4) only the left eye
    (5) only the right eye

46. Night blindness is *directly* caused by
    (1) lack of cones
    (2) lack of visual purple
    (3) lack of vitamin A
    (4) nerve impulses
    (5) nyctalopia

47. A cone is a
    (1) color-sensitive cell
    (2) frequency of light
    (3) nerve fiber
    (4) photochemical reaction
    (5) vitamin

48. Sunlight would stimulate
    (1) all of the cones
    (2) none of the cones
    (3) only the blue cones
    (4) only the red cones
    (5) only the yellow cones

The physical properties of different liquids can be used to separate mixed liquids. Liquids of different density and not soluble in each other segregate into layers. Liquids with different boiling points evaporate at different rates. Liquids can be classified as volatile and nonvolatile. Volatile liquids have low boiling points, like ethyl ether and carbon disulfide. Nonvolatile liquids have high boiling points; for example, mercury and sulfuric acid are relatively nonvolatile.

Liquids also vary in viscosity and surface tension. Viscosity reflects the relative ease of flow of a liquid. A highly viscous liquid flows slowly. Surface tension measures the attraction between surface molecules of a liquid. Liquid molecules below the surface of a liquid have attractive forces exerted on them from surrounding molecules in all directions, while molecules at the surface have an imbalance of forces causing them to cohere more closely together to form a surface film. The outermost layer of molecules, or film, can support objects floated on its surface even though the objects may have a higher density than the liquid.

49. Which property could be used to separate a solution of alcohol and water into its two components?
    (1) density
    (2) flammability
    (3) surface tension
    (4) viscosity
    (5) volatility

50. To have two liquids stirred together then separate as distinct layers, the two liquids must be
    (1) heated until one boils
    (2) highly viscous
    (3) the same density
    (4) under high pressure
    (5) not soluble in each other

51. Which liquid would have the highest boiling point?
    (1) carbon disulfide
    (2) ether
    (3) ethyl alcohol
    (4) mercury
    (5) water

52. Bugs that walk atop a pond take advantage of water's
    (1) boiling point
    (2) density
    (3) surface tension
    (4) viscosity
    (5) volatility

Stratigraphy is the science of layered rocks. It is based on the law of superposition, which states that the beds occur in the sequence as they were deposited, the oldest at the bottom and the youngest at the top, unless they were disturbed by later earth movements. The basic stratigraphic unit is the formation, which has a distinctive appearance and is thick enough to be mapped. Formations are named after a type locality, which is where they were first described. Thus, the famous St. Peter Sandstone was named for a town in Minnesota. Formations may be divided into members or combined into groups. Members and groups are also given geographical names.

The boundary between two beds (strata) is called the bedding plane. The formations of a given locality can be charted as a columnar section, in which the older rock is drawn beneath the younger one as in the normal sequence in the earth. To plot the normal sequence of strata, it is necessary to know which is the top of a sedimentary bed and which is the bottom. Many means are available to indicate the original position of sedimentary rocks. The beds were originally laid down nearly flat, and the upper surface often received such features as animal tracks, ripple marks, mud cracks, and raindrop prints. These features show which side of the rock was up and which was down, for they have a different appearance from above than from below.

53. Five formations are labeled in the cross-section through 3 hills shown in figure 38. Use the law of superposition to determine which is the youngest.
    (1) A   (2) B   (3) C   (4) D   (5) E

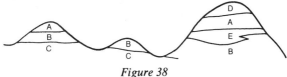

Figure 38

54. Which of these properties does *not* necessarily apply to a formation?
    (1) recognizable
    (2) geographical name
    (3) mappable
    (4) rock unit
    (5) very hard

55. The correct sequence of stratigraphic units, from smallest to largest, is
    (1) formation, group, member
    (2) group, formation, member
    (3) group, member, formation
    (4) member, formation, group
    (5) member, group, formation

56. Which feature is *not* useful for finding the original top side of a sedimentary bed?
    (1) animal footprints
    (2) color pattern
    (3) mud cracks
    (4) raindrop impressions
    (5) ripple marks

Viruses are intracellular parasites living within either the cytoplasm or the nucleus of the host cell. They are classified by Bergey in the order Virales in the class Microtatobiotes, but many microbiologists do not consider them living organisms. They have no cellular organization. They consist of protein and nucleic acid, either DNA or RNA; some also contain lipids or carbohydrates. They have few or no enzymes and rely on the metabolism of their hosts for energy and growth. Viruses may be crystallized. In the crystalline state they are similar to inanimate chemical crystals and do not appear to be living, but they retain their infectious ability. Characteristics they share with living things include reproduction (within host cells) and the possession of genetic material. With a few exceptions, viruses are submicroscopic, having dimensions from 10 to 300 nanometers. They can be observed with the electron microscope. They are spherical, polyhedral, rod-shaped, or tadpole-shaped. Most viruses pass through ordinary bacteriological filters made of porcelain or diatomaceous earth, but they are retained by collodion or cellulose acetate filters, called membrane filters. Viruses are parasites of vertebrates, invertebrates, plants, and bacteria. Some viruses produce diseases in their hosts, but others seem to be harmless. Generally, a cell containing a virus is immune to infections by other viruses of the same type, but not to other viruses.

57. Which of the following features suggests that viruses are alive?
    (1) ability to be crystallized
    (2) capacity to reproduce
    (3) cellular organization
    (4) efficient metabolism
    (5) polyhedral shape

58. What genetic material do viruses possess?
    (1) carbohydrates
    (2) cytoplasm
    (3) enzymes
    (4) nucleic acid
    (5) protein

59. Which property of viruses explains why bacteriological filters are unable to retain them?
    (1) their inanimate habit
    (2) their parasitic habit
    (3) their size
    (4) their shapes
    (5) they eat porcelain and diatomaceous earth

60. Which of these objects could *not* be host for a virus?
    (1) carbohydrate
    (2) horse
    (3) lobster
    (4) *Streptococcus*
    (5) yellow pine

ANSWER KEY FOR PRACTICE TEST 1

The Answer Key has been coded to assist you in identifying subjects for review.

B—Biology
G—Geology
C—Chemistry
P—Physics

1. (3) C	26. (1) P	51. (4) P
2. (4) B	27. (3) P	52. (3) P
3. (5) B	28. (4) P	53. (4) G
4. (2) C	29. (1) G	54. (5) G
5. (4) B	30. (4) G	55. (4) G
6. (5) B	31. (3) G	56. (2) G
7. (5) P	32. (5) G	57. (2) B
8. (5) B	33. (2) B	58. (4) B
9. (3) G	34. (1) B	59. (3) B
10. (2) C	35. (4) B	60. (1) B
11. (2) G	36. (5) B	
12. (2) P	37. (5) C	
13. (5) B	38. (3) C	
14. (5) P	39. (1) C	
15. (3) C	40. (3) C	
16. (4) G	41. (5) G	
17. (4) P	42. (5) G	
18. (4) B	43. (5) G	
19. (3) B	44. (2) G	
20. (4) C	45. (1) B	
21. (2) B	46. (2) B	
22. (5) B	47. (1) B	
23. (3) B	48. (1) B	
24. (5) B	49. (5) P	
25. (1) P	50. (5) P	

SCORING YOUR GED SCIENCE TEST

To score your GED Science Test total the number of correct answers for the test. Do not subtract any points for questions attempted but missed, as there is no penalty for guessing. This score is scaled from 20 to 80. GED score requirements vary from state to state.

PRACTICE TEST 1: SCORE APPROXIMATOR

The following Score Approximator is designed to assist you in evaluating your skills and to give you a very general indication of your scoring potential.

Number of Correct Answers	Approximate GED Score
50–60	63–80
40–49	54–62
30–39	47–53
20–29	40–46
10–19	26–39
1–9	20–25

ANALYZING YOUR TEST RESULTS

The charts on the following pages should be used to carefully analyze your results and spot your strengths and weaknesses. The complete process of analyzing each subject area and each individual question should be completed for each practice test. These results should be reexamined for trends in types of errors (repeated errors) or poor results in specific subject areas. THIS REEXAMINATION AND ANALYSIS IS OF TREMENDOUS IMPORTANCE TO YOU IN ASSURING MAXIMUM TEST PREPARATION BENEFIT.

PRACTICE TEST 1: GENERAL ANALYSIS SHEET

	Possible	Completed	Right	Wrong
Part A Background Knowledge	20			
Part B Reading Comprehension	40			
TOTAL	60			

PRACTICE TEST 1: SUBJECT AREA ANALYSIS SHEET

The coded Answer Key will assist you in this analysis.

	Possible	Completed	Right	Wrong
Biology	24			
Geology	15			
Chemistry	9			
Physics	12			
TOTAL	60			

ANALYSIS—TALLY SHEET FOR QUESTIONS MISSED

One of the most important parts of test preparation is analyzing WHY you missed a question so that you can reduce the number of mistakes. Now that you have taken Practice Test 1 and corrected your answers, carefully tally your mistakes by marking them in the proper column.

REASON FOR MISTAKE

	Total Missed	Simple Mistake	Misread Problem	Lack of Knowledge
Biology				
Geology				
Chemistry				
Physics				
TOTAL				

Reviewing the above data should help you determine WHY you are missing certain questions. Now that you have pinpointed the type of error, when you take Practice Test 2 focus on avoiding your most common type.

COMPLETE ANSWERS AND EXPLANATIONS FOR PRACTICE TEST 1

Part A: General Science Knowledge

1. (3) Gas expands at higher temperature and contracts at higher pressure. The relations are known as Charles' law and Boyle's law, respectively.

2. (4) Protozoans are one-celled creatures, and cells characteristically contain (among other things) cytoplasm (the main cell filling), plastids (metabolic bodies), chromosomes (genetic material), and vacuoles (voids). Stamens, however, are parts of flowers, the organs with pollen.

3. (5) Three familiar kinds of fungi are mushrooms, molds, and yeasts. Pizza is often flavored with mushrooms. One mold generates the penicillin antibiotics. Yeasts are used to brew beer and to make dough rise. On the other hand, milk is pasteurized by heating to kill bacteria.

4. (2) Avogadro's number, about 6×10^{23}, is the number of atoms in a standard weight of a chemical element, the grams equal to its atomic weight; 16 grams of oxygen, with an atomic weight of 16, would contain 6×10^{23} atoms.

5. (4) Enzymes are special proteins which catalyze (speed up) specific biochemical reactions within an organism, like the digestion of fats. The secretions which coordinate the body by controlling the organs are called hormones.

6. (5) A tropism is the movement of a plant due to a stimulus, like sunlight. Conditioned reflexes apply to the trained behavior of animals. Refraction is the bending of light rays from one transparent material to another, as from air to water.

7. (5) Gravitational attraction by the moon (and to a much lesser extent, the sun) causes tides, as the moon's mass attracts and distorts the earth and its oceans. You may remember that the height of tides corresponds to the lunar phase. The highest tides (spring tides) occur at new and full moon, when the sun's attraction is aligned with the moon's.

8. (5) Natural selection was Darwin's proposed mechanism for the evolution (or change) of life forms. Varieties that were able to compete more successfully for food would reproduce more frequently than less able varieties, and so a species would slowly change toward its more successful variations. Small changes cumulated over millions of years could lead to strikingly new forms of life.

9. (3) The probes showed Mars to be cratered rather like the earth's moon. The two satellites of Mars were discovered by telescope in 1877. There is no evidence of canals on Mars, and life on that planet has not been discovered, only conjectured. The Voyager space probes found erupting volcanoes on a moon of Jupiter.

10. (2) By definition, all acids contain hydrogen. Acidity is measured by the concentration of hydrogen ions in solution. Hydrochloric acid is HCl and sulfuric acid is H_2SO_4.

11. (2) The only city in the Southern Hemisphere is Buenos Aires, Argentina. July is wintertime there and hence colder than any of the northern cities. The 23° tilt of the earth's axis causes the seasons, with contrasting seasons on either side of the equator.

12. (2) You should know that a kilometer is shorter than a mile, being about ⅝ of one mile. Therefore a mile is about ⅘ of one kilometer. So 100 km/hr times 0.62 mi/km equals 62 mi/hr.

13. (5) Two species living together intimately is the definition of symbiosis. Embolism is a circulatory problem due to a blood clot. Metabolism is the set of chemical processes by which life obtains and uses energy. Mitosis is cell splitting, while osmosis is absorption.

14. (5) Fusion of hydrogen nuclei (as in the hydrogen bomb) is the source of the sun's energy. Because the sun is composed mostly of hydrogen, it has radiated and will continue to radiate for billions of years. The sun contains almost no heavy elements, and so the fission of uranium is impossible. No chemical reaction, including combustion, could sustain the sun's energy output.

15. (3) The valence number determines the number of other atoms with which an atom of a particular element can combine. For example, oxygen has a valence of two and combines with two hydrogen atoms (water, H_2O), hydrogen having a valence of one. Carbon of valence four would combine with two oxygens (carbon dioxide, CO_2) or four hydrogens (methane, CH_4).

16. (4) Nitrogen is obtained from the atmosphere, oxygen and nitrogen being separated during the liquefaction of air. The other four elements are metals obtained from mines.

17. (4) Electricity, microwaves, radio waves, and light are electromagnetic phenomena, differing in wavelength and frequency, but all traveling at the speed of light, 186,000 miles per second. Sound waves are pressure waves in air or another medium, and you should know from jet aviation that the sonic velocity in air is about 740 miles an hour.

18. (4) Photosynthesis is the process by which green plants produce sugar, the reaction being:

$$6CO_2 + 6H_2O \xrightarrow{\text{light and chlorophyll}} C_6H_{12}O_6 + 6O_2$$

carbon dioxide   *water*   *sugar*   *oxygen*

Remember that plants take in carbon dioxide and give off oxygen.

19. (3) An earthworm is in the annelid phylum of segmented worms. Insects, spiders, and crustaceans (crab, shrimp) are arthropods.

20. (4) The strength of an acid or base is reported as a pH number. A neutral solution has a pH of 7, acids are less than 7, and bases are greater than 7.

Part B: Science Reading Comprehension

21. (2) In photosynthesis, plants take in carbon dioxide and give off oxygen. But photosynthesis does not involve nitrogen.

22. (5) The decay of hydrocarbons in dead plants and animals is a process of oxidation, taking oxygen out of the air and producing carbon dioxide. At the same time, ammonia (NH_3) is released. Thus all three cycles are involved in decay.

23. (3) After decay and nitrifying bacteria have produced nitrates, the nitrogen is available for protein. Finally *denitrifying microorganisms* decompose nitrates into nitrogen gas.

24. (5) CO_2, sunlight, and water are essential for photosynthesis. Nitrogen is another plant nutrient and it is made available by nitrifying bacteria. However, the function of the denitrifying bacteria is to convert nitrates to nitrogen gas, which is not usable by plants.

25. (1) The electromagnetic field of an accelerator acts on the electrical charge of the particles. Note that neutrons cannot be accelerated because they are uncharged.

26. (1) The betatron is suitable for electrons only.

27. (3) The alpha particle is a helium nucleus, with 2 protons and 4 nucleons (protons and neutrons). The passage states that the copper is bombarded with an alpha particle, which therefore must be on the left side of the reaction.

28. (4) For gallium, the subscript 31 is its atomic number and its number of protons. The superscript 66 is the atomic mass and its number of nucleons (protons and neutrons). So the number of neutrons = 66 − 31 = 35.

29. (1) Fossils are the record of ancient life. Preserved worm burrows, for instance, record the existence of ancient worms, and so the burrows are fossils. But cave stalactites are inorganic dripstone produced by deposition in a cavern. Prehistoric tools are evidences of early man.

30. (4) Organic remains are petrified or preserved when circulating subsurface solutions deposit mineral matter in the open spaces within the remains. After the organic material has disappeared, the mineral matter retains details of the shape of the organism.

31. (3) Sand dunes were accumulated by the action of the wind, whether in the desert or on a beach. Dune sands were then exposed to air, which is essential to most decay bacteria.

32. (5) The durable parts of animals are most likely to be preserved. Because a worm has only soft parts, it usually decays and is rarely fossilized.

33. (2) The passage describes the oxygen as being transferred from the air through the lungs into the blood, which carries it to the tissues. Carbon dioxide travels in the reverse sequence.

34. (1) According to the passage, carbon dioxide is carried as dissolved sodium bicarbonate. (Later work suggests that this idea is incorrect, most of the CO_2 being transported as carbonic acid within the red blood cells.)

35. (4) Each gas diffuses due to pressure differences for that particular gas. So the CO_2 pressure is lower in the alveoli than in the venous blood.

36. (5) Carbon dioxide is waste generated during metabolism of the tissue cells. It diffuses into the veins, where bicarbonate forms to permit its passage to the lungs.

37. (5) Partial combustion of carbon yields CO, while complete combustion produces CO_2. The latter contains more oxygen in combination with the carbon.

38. (3) Charcoal is nearly pure carbon, and therefore its combustion yields oxides of carbon. The brisk wind would supply enough oxygen for complete oxidation to carbon dioxide.

39. (1) In any chemical reaction, substances to the left of the arrow (reactants) are transformed to those to the right (products). An enzyme is an organic catalyst, which facilitates the reaction without itself being changed. A catalyst is not a reactant.

40. (3) Dry ice is the solid form of carbon dioxide, which is a gas at normal temperature and pressure.

41. (5) Corrasion is wind erosion, which would be suppressed by barren rock (because wind moves small particles), forest (which breaks the wind and anchors the soil), high rainfall (which fosters vegetation), or lack of wind. But low elevation would not prevent corrasion. Death Valley, California, has the lowest altitude in the United States and appreciable wind erosion.

42. (5) The largest particles movable by wind simply roll or slide under the impact of smaller particles; the process is called traction. Note that beveled stones called ventifacts may be fixed in place rather than be moved.

43. (5) The passage states that wind is a very selective agent of transportation, meaning it readily separates material of different sizes. Gravel is too coarse to be easily moved by wind. Dunes tend to be produced, rather than leveled, by wind.

44. (2) Saltation, or bouncing, is the chief means by which wind transports medium-size (sand-size) particles.

45. (1) Because the crossover of optic nerve fibers is 50 percent, each eye sends the same amount of information to each half of the brain.

46. (2) Night blindness or nyctalopia is caused by a lack of visual purple, which may be attributed to a deficiency of vitamin A. *Indirectly,* then, does a lack of the vitamin induce night blindness.

47. (1) A cone is a cell sensitive to a particular color of light.

48. (1) Because sunlight is nearly white, it is compounded from all frequencies (colors) of light. This light would stimulate the red, blue, and green cones approximately equally. Note that there are no "yellow cones."

49. (5) The distillation of an alcohol-water solution is based on their distinct volatilities. Heating the solution first boils off the alcohol, which has a lower boiling point than water.

50. (5) The passage states that a liquid mixture will separate to different layers if the two liquids were mutually insoluble (immiscible) and of different densities. Thus an oil-vinegar mixture like salad dressing divides to a layer of lighter oil floating atop the heavier vinegar.

51. (4) Mercury is given as an example of a nonvolatile liquid, implying that it has a relatively high boiling point. Mercury boils at 357°C, far higher than water or the other volatile liquids mentioned.

52. (3) Bugs can walk on the pond's surface because the surface tension of the water supports them.

53. (4) Formation D is the youngest shown, for the law of superposition states that younger formations are deposited atop older formations.

54. (5) A formation is a recognizable, mappable rock unit which is given a geographical name. The rock type may be either hard (as quartzite) or soft (as shale). The term formation does not imply hardening or metamorphism.

55. (4) A member is a part of a formation, which is a part of a group.

56. (2) The passage says nothing about color pattern, though the other four features are listed as of use in interpreting original orientation.

57. (2) Their ability to reproduce suggests that they are alive. They lack cellular organization and metabolic ability. The polyhedral (geometric) shape of some viruses and their crystallizability suggest that they are inanimate chemicals.

58. (4) The genetic material within viruses is either ribose nucleic acid (RNA) or deoxyribose nucleic acid (DNA). A virus requires the host cell for its reproduction.

59. (3) Viruses are so very small that they pass through ordinary filters.

60. (1) A virus can infect animals (horse, lobster), plants (pine), or bacteria (*Streptococcus* is the microorganism that causes "strep throat"). A carbohydrate is not an organism, only an organic chemical.

BEFORE YOU TAKE THE NEXT PRACTICE TEST
1. Reread the ANALYSIS OF EXAM AREAS (page 9).
2. Skim the 4 science review sections (page 37).
3. Reread the explanations for all questions you missed.
4. Remember to use the SYSTEMATIC OVERALL APPROACH.

PRACTICE TEST 2
WITH COMPLETE ANSWERS AND EXPLANATIONS

ANSWER SHEET FOR PRACTICE TEST 2
(Remove This Sheet and Use It to Mark Your Answers)

1 ① ② ③ ④ ⑤	26 ① ② ③ ④ ⑤	51 ① ② ③ ④ ⑤
2 ① ② ③ ④ ⑤	27 ① ② ③ ④ ⑤	52 ① ② ③ ④ ⑤
3 ① ② ③ ④ ⑤	28 ① ② ③ ④ ⑤	53 ① ② ③ ④ ⑤
4 ① ② ③ ④ ⑤	29 ① ② ③ ④ ⑤	54 ① ② ③ ④ ⑤
5 ① ② ③ ④ ⑤	30 ① ② ③ ④ ⑤	55 ① ② ③ ④ ⑤
6 ① ② ③ ④ ⑤	31 ① ② ③ ④ ⑤	56 ① ② ③ ④ ⑤
7 ① ② ③ ④ ⑤	32 ① ② ③ ④ ⑤	57 ① ② ③ ④ ⑤
8 ① ② ③ ④ ⑤	33 ① ② ③ ④ ⑤	58 ① ② ③ ④ ⑤
9 ① ② ③ ④ ⑤	34 ① ② ③ ④ ⑤	59 ① ② ③ ④ ⑤
10 ① ② ③ ④ ⑤	35 ① ② ③ ④ ⑤	60 ① ② ③ ④ ⑤
11 ① ② ③ ④ ⑤	36 ① ② ③ ④ ⑤	
12 ① ② ③ ④ ⑤	37 ① ② ③ ④ ⑤	
13 ① ② ③ ④ ⑤	38 ① ② ③ ④ ⑤	
14 ① ② ③ ④ ⑤	39 ① ② ③ ④ ⑤	
15 ① ② ③ ④ ⑤	40 ① ② ③ ④ ⑤	
16 ① ② ③ ④ ⑤	41 ① ② ③ ④ ⑤	
17 ① ② ③ ④ ⑤	42 ① ② ③ ④ ⑤	
18 ① ② ③ ④ ⑤	43 ① ② ③ ④ ⑤	
19 ① ② ③ ④ ⑤	44 ① ② ③ ④ ⑤	
20 ① ② ③ ④ ⑤	45 ① ② ③ ④ ⑤	
21 ① ② ③ ④ ⑤	46 ① ② ③ ④ ⑤	
22 ① ② ③ ④ ⑤	47 ① ② ③ ④ ⑤	
23 ① ② ③ ④ ⑤	48 ① ② ③ ④ ⑤	
24 ① ② ③ ④ ⑤	49 ① ② ③ ④ ⑤	
25 ① ② ③ ④ ⑤	50 ① ② ③ ④ ⑤	

CUT HERE

PRACTICE TEST 2

Time: 90 Minutes
60 Questions

DIRECTIONS

The Science Test consists of 60 multiple-choice questions in two parts. In the first part are 20 questions based on background knowledge of scientific concepts. The second part contains 40 questions based on several brief reading passages. You have 90 minutes to complete both parts of this test. There is no penalty for guessing.

Part A: General Science Knowledge

The following questions are not based on a reading passage. Select the best answer based on your knowledge of the concepts of natural science.

1. Oysters and clams belong to the phylum of
    (1) arthropods
    (2) chordates
    (3) coelenterates
    (4) echinoderms
    (5) mollusks

2. Diabetes mellitus, or "sugar diabetes," is due to malfunction of the
    (1) bone marrow
    (2) kidneys
    (3) pancreas
    (4) stomach
    (5) thyroid gland

3. Parallax has been used to measure the
    (1) composition of many planets
    (2) distance to some stars
    (3) length of the sidereal year
    (4) likelihood of solar flares
    (5) temperature of the moon

4. The group of elements lithium, sodium, and potassium would be described as
    (1) highly reactive gases
    (2) highly reactive metals
    (3) slightly reactive metals
    (4) slightly reactive salts
    (5) unreactive gases

5. Two isotopes of one chemical element differ in their number of
   (1) electrons
   (2) neutrons
   (3) protons
   (4) electrons and protons
   (5) both neutrons and protons

6. Granite forms by
   (1) accumulation in deltas
   (2) compression of plant remains
   (3) crystallization of a melt
   (4) heating of limestone
   (5) precipitation from the ocean

7. Which *botanical community* would be found furthest north?
   (1) coniferous forest
   (2) deciduous forest
   (3) grassland
   (4) permafrost
   (5) tundra

8. A temperature of 230° Fahrenheit equals
   (1) 80°C
   (2) 90°C
   (3) 93°C
   (4) 100°C
   (5) 110°C

9. When a cross of red flowers with white flowers leads to a next generation of only white flowers, the likeliest explanation is that
   (1) insects are attracted to red plants
   (2) mutation has occurred
   (3) sunlight bleached the flowers
   (4) the red color is recessive
   (5) the soil lacked iron

10. The characteristic chemical element within organic compounds is
    (1) carbon
    (2) nitrogen
    (3) oxygen
    (4) protein
    (5) sulfur

11. Glaciers demonstrate that
    (1) rivers freeze during wintertime
    (2) solids can flow slowly under pressure
    (3) the earth's interior is cold
    (4) the earth is colder than in the past
    (5) the Ice Age is returning

12. Halley's Comet has been shown to move in a path that is
    (1) circular
    (2) elliptical
    (3) irregular
    (4) linear
    (5) parabolic

13. As ice cakes floating on a pond of water *begin* to melt in springtime, the water temperature
    (1) exceeds 32°F
    (2) falls
    (3) rises
    (4) remains the same
    (5) is indeterminate

14. Gametes are the direct consequence of
    (1) meiosis
    (2) mitosis
    (3) mutation
    (4) osmosis
    (5) selection

15. As magnesium fluoride dissolves in water, it dissociates to
    (1) atoms
    (2) ions
    (3) metals
    (4) molecules
    (5) salts

16. Oceanic dredging is unlikely to recover
    (1) coral atolls
    (2) diatoms
    (3) limestone
    (4) manganese ore
    (5) volcanic rocks

17. The basic building blocks of proteins are
    (1) amino acids
    (2) carbohydrates
    (3) minerals
    (4) phosphates
    (5) vitamins

18. When completely immersed in water, a piece of glass
    (1) appears bent by refraction
    (2) appears to weigh less
    (3) decreases in density
    (4) increases in density
    (5) increases its volume

19. The similar arrangement of bones within a man's arm and a whale's flipper suggests that
    (1) coincidences can occur
    (2) man evolved from fishes
    (3) men are meant to swim
    (4) men and whales have a common ancestor
    (5) similar tasks have led to similar structures

20. The function of deoxyribose nucleic acid is to
    (1) bond protons and neutrons
    (2) digest food
    (3) neutralize alkaline solutions
    (4) prevent precipitation of salt in the oceans
    (5) store genetic information

Part B: Science Reading Comprehension

This section contains 10 reading passages, with several questions about each passage. Read the passage first and then answer the questions following it. Refer to the passage as often as necessary in selecting the best answer. Remember, there is no penalty for guessing.

Tuberculosis is a chronic disease endemic in many countries. It commonly infects the lungs, where the invading cells are usually isolated in tubercles, or nodules, formed by the host as a reaction to the pathogen. While the tubercles remain intact, the bacteria do little harm, and the host may acquire a degree of immunity. If host resistance is low, growth of bacteria within the tubercle may cause it to burst, with resulting spread of bacteria to healthy tissue. Advanced cases of tuberculosis are characterized by the coughing up of highly infective sputum containing dead lung tissue and tubercle bacilli. The latter are highly resistant and can remain viable for long periods in dry sputum. Excessive lung damage causes death. *Mycobacterium tuberculosis* is pathogenic for cattle as well as man and may be spread from cows to man through milk. Organisms entering the human body in milk usually cause tuberculosis of the bone. Control methods include isolation of patients, destruction of tubercular cattle, and pasteurization of milk. The BCG vaccine has been used with some success in Europe, but it is not in general use in the United States. A positive reaction to the tuberculin test indicates a present or previous infection. X rays reveal tubercles in the lung. Presence of *M. tuberculosis* in sputum indicates an active infection.

21. Which is *not* part of the description of tuberculosis?
    (1) bacterial
    (2) chronic
    (3) hereditary
    (4) infection
    (5) respiratory

22. Death from tuberculosis is *directly* ascribed to
    (1) bacilli
    (2) BCG
    (3) lung damage
    (4) sputum
    (5) tubercles

23. What is the genus of the tubercular microorganism?
    (1) bacilli
    (2) *Mycobacterium*
    (3) endemic
    (4) pathogen
    (5) *tuberculosis*

24. All of the following are mentioned as weapons against tuberculosis *except*
    (1) isolation
    (2) pasteurization
    (3) vaccination
    (4) vitamins
    (5) X rays

    A volcano emits solid, liquid, and gaseous matter. According to their size, solid fragments (pyroclastic material) thrown out by a volcano are called volcanic dust, ash, cinders, blocks (if angular), or bombs (if rounded). Liquid lava may be smooth and ropy (called pahoehoe) or rough and porous (called aa). Most volcanic gas is steam, but carbon dioxide is common, and compounds of sulfur, hydrogen, chlorine, and fluorine are usually detectable. Much volcanic rock is believed to have been deposited by so-called fiery clouds, which carry a slurry of hot particles and gases downslope at speeds of perhaps 60 miles per hour.

    The nature of an eruption, as well as the type of cone produced, depends mostly on the composition of the eruption. An abundance of gas causes violent eruptions (such as Krakatoa) and accumulation of much pyroclastic debris. A minimum of gas results in relatively quiet lava flows, as in Hawaii. Volcanoes may change their style of eruption and alternate between quiet and explosive activity, as Vesuvius has done.

    Cinder cones consist of medium-sized particles, which are able to stand at fairly steep angles. Lava cones or shield volcanoes are formed by successive flows of lava and are gentle in slope, like the profile of a warrior's shield on the ground. Stratovolcanoes have composite cones of alternating layers of fragments and lava.

25. According to the article, cinders must be
    (1) gaseous matter
    (2) pyroclastic fragments
    (3) rounded
    (4) finer than ash particles
    (5) lava flows

26. A volcano with plentiful steam and carbon dioxide is likely to
    (1) be a shield volcano
    (2) be unusually large
    (3) emit pahoehoe lava
    (4) erupt violently
    (5) have no hydrogen sulfide (H_2S) odor

27. Which material is most likely to have been deposited from a so-called fiery cloud?
    (1) aa
    (2) ash
    (3) blocks
    (4) bombs
    (5) pahoehoe

28. Which volcano is matched correctly with its type?
    (1) Hawaii and cinder cone
    (2) Hawaii and stratovolcano
    (3) Krakatoa and shield volcano
    (4) Vesuvius and shield volcano
    (5) Vesuvius and stratovolcano

To produce sound, there must be present both a vibrating source that initiates a mechanical disturbance (wave) and an elastic medium through which the wave can be transmitted. Consider a simple experiment to demonstrate the need for an elastic substance to carry the sound. If an electric buzzer is hung inside a bell jar so that it does not touch the sides of the jar, the sound of the buzzer can be heard when air is inside the jar, because the air transmits the sound waves. As soon as the bell jar is exhausted by a vacuum pump, the sound can no longer be heard because there is no material through which the disturbance can travel. By tilting the evacuated bell jar so that the buzzer touches the wall of the jar, the sound can once again be heard; therefore a solid (the glass of the jar) can carry the sound wave as well as a gas (the initial air). In a second experiment, you could show that a liquid, too, can transmit sound by ringing a small bell beneath the surface of water in a sink or large pan.

29. It would be most correct to say that sound travels as a(n)
    (1) buzzing
    (2) disturbance
    (3) electric current
    (4) gas
    (5) vacuum

30. In the first experiment, the sound wave originates from the
    (1) air
    (2) bell
    (3) buzzer
    (4) glass
    (5) vacuum

31. When the pumping begins in the first experiment, the buzzing heard by the observer should
    (1) abruptly stop
    (2) become louder
    (3) become louder, then diminish
    (4) diminish at first, then become louder
    (5) gradually diminish

32. In the second experiment, the *elastic medium* through which the sound wave is transmitted is the
    (1) air
    (2) bell
    (3) glass
    (4) sink or pan
    (5) water

    The crayfish (or crawfish) is a freshwater crustacean commonly studied as a representative arthropod. Living crayfish are brown or greenish in color, but after death they become red.
    The crayfish body has two main regions, the cephalothorax and the abdomen. The exoskeleton of the cephalothorax consists of a shieldlike structure called the carapace. The only evidence of segmentation on the carapace is the cervical groove, which marks the junction of the head and thorax. On the ventral surface of the cephalothorax are a variety of paired appendages: two pairs of antennae, one larger than the other; mouth parts consisting of a pair of mandibles, two pairs of maxillae, and three pairs of maxillipeds; and five pairs of legs equipped with sensory hairs. The first pair of legs are the large pincers (chelipeds) used in obtaining food and for defense. The segmentation of the abdomen is obvious from the dorsal surface. It has six pairs of appendages, the swimmerets. In the male the first two pairs are modified for the transfer of sperm. The sixth pair (uropods) along with the terminal segment (telson) constitutes a fanlike tail used in swimming backward.

33. Where is the cephalothorax in relation to the abdomen?
    (1) above
    (2) behind
    (3) below
    (4) in front
    (5) it cannot be determined from the passage

34. The last segment of the abdomen is called the
    (1) carapace
    (2) maxilliped
    (3) pincer
    (4) telson
    (5) uropod

35. The chelipeds are structurally
    (1) an abdomen
    (2) antennae
    (3) a tail
    (4) legs
    (5) mandibles

36. The total number of appendages on a crayfish must be
    (1) 8   (2) 18   (3) 28   (4) 38   (5) 48

The elements fluorine, chlorine, bromine, and iodine are called the halogens, a name meaning salt-formers. Fluorine is commonly found in such minerals as fluorite (CaF_2) and cryolite (Na_3AlF_6). Chlorine occurs as various metal chlorides, the most familiar being common sodium chloride. Bromine, in the form of bromides, is found in natural salt brines. The most common mineral containing iodine is $NaIO_3$ in Chilean saltpeter.

Fluorine is made by the electrolysis of potassium fluoride (KF) in liquid hydrogen fluoride:

$$2KF \xrightarrow{HF} 2K + F_2$$

Chlorine is prepared industrially by the electrolysis of a sodium chloride brine in cells specially designed to keep the reactive anode and cathode products separate:

$$2NaCl + 2H_2O \xrightarrow{\text{electric current}} 2NaOH + H_2 + Cl_2$$

Although the other halogens may be prepared by electrolysis, they are usually produced by replacement by the more active chlorine:

$$2NaBr + Cl_2 \longrightarrow 2NaCl + Br_2$$

At room temperature, fluorine is a pale yellow gas, chlorine is a greenish gas, bromine is a red liquid, and iodine occurs as gray crystals. Some other physical properties of the halogens are given in this chart.

Element	Atomic Number	Melting Point	Boiling Point
fluorine	9	−219°C	−188°C
chlorine	17	−102°C	−35°C
bromine	35	−7°C	59°C
iodine	53	113°C	184°C

37. Common table salt is especially rich in which halogen?
    - (1) bromine
    - (2) chlorine
    - (3) fluorine
    - (4) iodine
    - (5) sodium

38. The halogen with the greatest temperature *range* in the liquid state is
    - (1) bromine
    - (2) chlorine
    - (3) brine
    - (4) fluorine
    - (5) iodine

39. Which halogen is a solid under normal conditions?
    - (1) bromine
    - (2) chlorine
    - (3) fluorine
    - (4) fluorite
    - (5) iodine

40. Considering the three reactions in the passage, which substance must be available for a procedure *before* bromine can be produced?
    - (1) cryolite
    - (2) hydrogen fluoride
    - (3) potassium fluoride
    - (4) saltpeter
    - (5) sodium chloride

Lakes may be fed by precipitation, stream water, groundwater, or melting snow. The geological problem of lakes has to do with the origin of lake basins, which may form in numerous ways. Warping of the land and faulting account for the origin of some lakes, often large ones. Glaciers scour out hollows on the surface or deposit their debris irregularly, thereby providing basins in which water accumulates. Karst terrain from the solution of limestone has many small lakes in sinkholes and collapsed caverns. Volcanoes may contain lakes in their craters. Wind erosion can also create depressions for lakes in arid regions.

Saline lakes are characteristic features of arid regions. Some have been isolated from the ocean, but most contain concentrations of salts dissolved from the surrounding land. Salt lakes are usually enriched in common salt, sodium chloride. Alkaline lakes contain potassium or sodium carbonate. Some lakes of unusual composition have large amounts of dissolved borax or other salts. A lake that expands and contracts seasonally, or dries up entirely at times, is a playa lake. Playa lakes may evaporate to leave valuable mineral deposits, which may be buried by later sediment.

Lakes are eventually destroyed by either being filled with sediment or drained of their water by downward erosion of the stream outlet from the basin. Vegetation, salt deposits, and deltas help to fill the lakes in which they form.

41. Lakes *cannot* originate by
    (1) evaporation
    (2) faulting
    (3) glaciation
    (4) volcanism
    (5) wind action

42. Most of the dissolved material in salty lakes comes from
    (1) erosion of rocks
    (2) evaporation
    (3) industrial pollution
    (4) the ocean
    (5) the wind

43. A minable deposit of borax is most likely on the site of an ancient
    (1) glacier
    (2) ocean
    (3) playa
    (4) sinkhole
    (5) swamp

44. Lakes are only temporary features of the landscape because
    (1) more sediment enters than leaves
    (2) precipitation is uncertain
    (3) they are isolated from the ocean
    (4) they evaporate
    (5) they fill up with salt

The menstrual cycle (see figure 39) is a period of regularly recurring changes which culminate in the discharge of blood and dead cells from the disintegration of the endometrium, a membrane inside the uterus. The menstrual cycle is initiated by the follicle-stimulating hormone (FSH) produced by the pituitary gland. FSH causes a few of the immature follicles to grow and release estrogens. A few days after the release of FSH, the pituitary starts putting out the luteinizing hormone (LH), which increases follicular growth and the production of estrogens. Finally, one of the follicles becomes so large that it ruptures, expelling its ovum (ovulation).

Following ovulation, the follicle cells increase in size and become

THE MENSTRUAL CYCLE

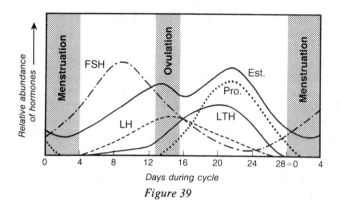

Figure 39

transformed into a yellow body called the corpus luteum. A third pituitary hormone, the leuteotropic hormone (LTH), stimulates the corpus luteum to secrete both estrogens and progesterone. Estrogens cause the endometrium to grow in thickness, and progesterone enhances endometrial blood flow and nutrient secretion. The output of LTH, LH, and FSH lasts for about 14 days, after which the endometrium disintegrates and releases blood and dead cells (menses). By convention, the cycle begins, and ends, on the first day the menses appear. The average cycle is of 28 days duration. Ovulation usually occurs on the fourteenth day, and menstruation lasts about four days.

45. The function of estrogen is to cause
    (1) follicular growth
    (2) ovulation
    (3) progesterone secretion
    (4) the lining of the uterus to grow in thickness
    (5) the pituitary gland to secrete FSH

46. Which of the hormones is (are) *not* present before ovulation?
    (1) estrogen
    (2) estrogen and LH
    (3) progesterone
    (4) progesterone and LTH
    (5) estrogen, progesterone, and LTH

47. Which hormone is secreted by both the follicles and the corpus luteum?
    (1) estrogen
    (2) FSH
    (3) LH
    (4) LTH
    (5) progesterone

48. The output of progesterone lasts a maximum of
    (1) 1 day
    (2) 4 days
    (3) 14 days
    (4) 17 days
    (5) 28 days

Because the property of electric charge exists in separate units rather than in any continuous amount, that property is said to be quantized. The concept of quantization is fundamental in modern physics. The existence of atoms and particles like electrons and protons demonstrates the quantization of matter. Energy, too, is a quantum phenomenon when examined on the atomic scale.

The unit, or quantum, of electric charge e is so tiny that the "graininess" of electricity is not displayed in large-scale experiments, just as we do not notice that the air we breathe is composed of separate molecules. In an ordinary light bulb, for instance, 10^{19} units of charge travel through the filament every second. Maxwell's classical electromagnetic theory does not mention the quantization of charge. Similarly, Newton's three laws of motion say nothing about the ultimate quantities of mass, the atomic and subatomic particles. Both of these great theories are incomplete in the sense that they fail to describe the behavior of electricity and matter on the atomic scale. The more detailed theories of quantum physics were devised to extend our knowledge to such phenomena.

49. Which of these large-scale measurements displays quantization?
    (1) altitude
    (2) area
    (3) population
    (4) rainfall
    (5) temperature

50. Maxwell's theory is referred to as *classical* because it
    (1) explains musical sounds
    (2) is outdated and incomplete
    (3) proved useful for a long time
    (4) was first stated by Aristotle
    (5) was not concerned with electric charge

51. Quantum physics has discovered the existence of
    (1) continuity
    (2) electric charge
    (3) graininess
    (4) mass
    (5) three laws of motion

52. The quantization of energy would imply that
    (1) electrons are restricted to specific levels of excitement
    (2) energy can be measured
    (3) experiments can approach absolute zero but not attain it
    (4) experiments would show that an automobile cannot travel at some speeds
    (5) particles of energy exist

Sulfur was known to man, as brimstone, throughout antiquity. Although the supply is dwindling, deposits of free (uncombined) sulfur are found in Texas and Louisiana. Sulfur is also found in the form of metal sulfides, like iron pyrite (FeS_2), chalcopyrite ($CuFeS_2$), and galena (PbS), and as sulfates like gypsum, $CaSO_4 \cdot 2H_2O$.

In volcanic regions, sulfur can be directly mined and then purified by heating. In areas of Texas and Louisiana, it is obtained by the Frasch process: superheated water and air are forced down pipes in boreholes to melt the sulfur and force it to the surface. The Frasch process makes use of four concentric pipes. The outer two pipes carry the superheated (about 180°C) water and the innermost pipe carries the compressed air. The molten sulfur is forced up through the third pipe.

Sulfur melts at 113°C and forms a straw-colored liquid called mobile sulfur. If the mobile sulfur is heated further, it darkens and becomes the thick, sticky viscous sulfur. If the viscous sulfur is suddenly cooled, it becomes the rubbery plastic sulfur. Plastic sulfur gradually changes back into crystalline sulfur under ordinary conditions.

53. Which substance contains the highest percentage of sulfur atoms?
    (1) gypsum
    (2) iron pyrite
    (3) chalcopyrite
    (4) galena
    (5) chalcopyrite and galena

54. The Frasch process is a practical method for obtaining sulfur because sulfur
    (1) becomes viscous when heated
    (2) can be mined in volcanic regions
    (3) is found in the form of metal sulfides
    (4) melts at a relatively low temperature
    (5) occurs in Texas and Louisiana

55. The composition of the deposits tapped by the Frasch process must be
    (1) $CaSO_4 \cdot 2H_2O$
    (2) $CuFeS_2$
    (3) FeS_2
    (4) PbS
    (5) S

56. The substance pumped down the fourth Frasch pipe serves to
    (1) dissolve the sulfur
    (2) heat the sulfur
    (3) heat the water
    (4) make the sulfur viscous
    (5) push the sulfur upward

Spore discharge in the ferns is performed by the annulus. The annulus is a single ring of cells that partially encircles the sporangium; it is interrupted by a few cells called lip cells that have thin walls. The cells of the annulus have thick walls on three sides: the inner side and the two sides that touch adjacent annulus cells. Other walls of these cells are thin, and through them water evaporates as the sporangium grows older. As the water content decreases, the cohesion of water molecules to each other and the adhesion of water molecules to the thick cell walls pull these walls together. The annulus cells shrink, the annulus shortens and straightens, and a tear begins to form at the weaker lip cells and continues across the base of the sporangium. The top of the sporangium, which has most of the spores in it, bends backward. As more water evaporates from the annulus cells, the tension on the remaining water becomes so great that the water is suddenly converted to water vapor. Since there is no longer any tension exerted by water on the cell walls, they quickly return to their original position. The annulus curves again, and the top of the sporangium snaps back in place, hurling its spores out like stones from a slingshot. This mechanism, which is found only among ferns, insures that spores will be disseminated in dry weather when they have the greatest chance of being carried a long distance.

57. The sporangium must be a
    (1) cell capable of giving rise to an entire plant
    (2) plant hormone which stimulates spore production
    (3) plant which produces sexual spores
    (4) single ring of spore cells
    (5) structure within which spores are produced

58. The lip cells apparently
    (1) are the outlet for water vapor
    (2) cause adhesion of water molecules
    (3) cause cohesion of water molecules
    (4) localize the site of tearing
    (5) spit out the spores

59. The "slingshot" is pulled taut by
    (1) ferns
    (2) shrinkage of the annulus
    (3) water vapor
    (4) wet weather
    (5) the top of the sporangium

60. What aspect of this mechanism assures that spores are discharged in dry weather?
    (1) adhesion of water to the cell walls
    (2) cohesion of water molecules to each other
    (3) bending of the sporangium
    (4) curving of the annulus
    (5) evaporation of water

ANSWER KEY FOR PRACTICE TEST 2

The Answer Key has been coded to assist you in identifying subjects for review.

B—Biology
G—Geology
C—Chemistry
P—Physics

1. (5) B
2. (3) B
3. (2) P
4. (2) C
5. (2) C
6. (3) G
7. (5) B
8. (5) P
9. (4) B
10. (1) C
11. (2) G
12. (2) P
13. (4) P
14. (1) B
15. (2) C
16. (1) G
17. (1) B
18. (2) P
19. (4) B
20. (5) B
21. (3) B
22. (3) B
23. (2) B
24. (4) B
25. (2) G
26. (4) G
27. (2) G
28. (5) G
29. (2) P
30. (3) P
31. (5) P
32. (5) P
33. (4) B
34. (4) B
35. (4) B
36. (4) B
37. (2) C
38. (5) C
39. (5) C
40. (5) C
41. (1) G
42. (1) G
43. (3) G
44. (1) G
45. (4) B
46. (3) B
47. (1) B
48. (4) B
49. (3) P
50. (3) P
51. (3) P
52. (1) P
53. (2) C
54. (4) C
55. (5) C
56. (5) C
57. (5) B
58. (4) B
59. (2) B
60. (5) B

SCORING YOUR GED SCIENCE TEST

To score your GED Science Test total the number of correct answers for the test. Do not subtract any points for questions attempted but missed, as there is no penalty for guessing. This score is scaled from 20 to 80. GED score requirements vary from state to state.

PRACTICE TEST 2: SCORE APPROXIMATOR

The following Score Approximator is designed to assist you in evaluating your skills and to give you a very general indication of your scoring potential.

Number of Correct Answers	Approximate GED Score
50–60	63–80
40–49	54–62
30–39	47–53
20–29	40–46
10–19	26–39
1–9	20–25

ANALYZING YOUR TEST RESULTS

The charts on the following pages should be used to carefully analyze your results and spot your strengths and weaknesses. The complete process of analyzing each subject area and each individual question should be completed for each practice test. These results should be reexamined for trends in types of errors (repeated errors) or poor results in specific subject areas. THIS REEXAMINATION AND ANALYSIS IS OF TREMENDOUS IMPORTANCE TO YOU IN ASSURING MAXIMUM TEST PREPARATION BENEFIT.

PRACTICE TEST 2: GENERAL ANALYSIS SHEET

	Possible	Completed	Right	Wrong
Part A Background Knowledge	20			
Part B Reading Comprehension	40			
TOTAL	60			

PRACTICE TEST 2: SUBJECT AREA ANALYSIS SHEET

The coded Answer Key will assist you in this analysis.

	Possible	Completed	Right	Wrong
Biology	24			
Geology	11			
Chemistry	12			
Physics	13			
TOTAL	60			

ANALYSIS—TALLY SHEET FOR QUESTIONS MISSED

One of the most important parts of test preparation is analyzing WHY you missed a question so that you can reduce the number of mistakes. Now that you have taken the Practice Test 2 and corrected your answers, carefully tally your mistakes by marking them in the proper column.

REASON FOR MISTAKE

	Total Missed	Simple Mistake	Misread Problem	Lack of Knowledge
Biology				
Geology				
Chemistry				
Physics				
TOTAL				

Reviewing the above data should help you determine WHY you are missing certain questions. Now that you have pinpointed the type of error, focus on avoiding your most common type.

COMPLETE ANSWERS AND EXPLANATIONS FOR PRACTICE TEST 2

Part A: General Science Knowledge

1. (5) The most important classes of mollusks are cephalopods (squids and octopuses), gastropods (snails), and pelecypods (clams and oysters). Arthropods include crustaceans and insects. Chordates are animals with backbones, from fish to mammals. Coelenterates are corals. Examples of echinoderms are sea urchins, sand dollars, and starfishes, all with the characteristic five-rayed symmetry.

2. (3) The disease of diabetes is a metabolic disorder resulting in excessive amounts of sugar in the blood plasma. The pancreas gland should secrete the hormone insulin to control the sugar level. Diabetics with an abnormal pancreas need extra insulin to counteract their symptoms.

3. (2) Parallax is a range-finding technique used to measure the distance to some nearby stars from the annual angular displacement of a nearby star against the background of more distant, relatively fixed stars. Behold parallax by noting the apparent position of a pencil in front of your face with only your right eye, then your left eye.

4. (2) They are highly reactive metals, the alkali metals of the first column of the periodic table. Other alkali metals include rubidium and cesium. They combine with reactive nonmetals to form salts. Two highly reactive nonmetals are fluorine and chlorine, both gases. The unreactive "noble" gases are helium, neon, argon, krypton, and xenon.

5. (2) Chemical elements are defined by their number of protons, a constant for each element. Some variation in the number of neutrons leads to different weight isotopes of the same element. For example, the eighth element of the periodic table, oxygen, has 8 protons in its nucleus. Most oxygen atoms also have 8 neutrons in the nucleus, giving the O^{16} isotope of atomic weight 16. The rarer O^{18} isotope has 10 neutrons in addition to the essential 8 protons.

6. (3) Granite forms as molten rock cools and crystallizes deep within the earth. The parent melt is called magma. Accumulations of mud and sand in the delta at a river mouth could lead to shale and sandstone. Limestone forms by precipitation of calcium carbonate from the ocean. Heating of limestone produces marble, and compression of plant debris leads to coal.

ANSWERS AND EXPLANATIONS FOR PRACTICE TEST 2   155

7. (5) Tundra is the sparse shrub-and-moss vegetation of high altitudes or the far north. Southward are the successive communities of conifers (evergreens), deciduous trees (which lose their leaves for the winter), and grasses. Permafrost is permanently frozen soil of the far north, not a community of plants.

8. (5) 230°F exceeds the boiling point of water, 212°F. In the choices, the sole temperature over the 100°C boiling point of water is 110°C. The precise conversion equation is: $C = 5/9(F - 32) = 5/9(198) = 110$.

9. (4) Probably the whiteness is dominant in a gene and the redness is recessive, masked by the whiteness. Gregor Mendel's pioneer crossing of smooth and wrinkled garden peas led to an all-smooth generation because the wrinkled character was recessive with respect to the dominant smoothness. Mutation is not a reasonable explanation because several flowers of each color must have been crossed, and it is improbable the rare mutation could have occurred in every fertile seed.

10. (1) Organic compounds have carbon frameworks. For example, simple sugar from many fruits (dextrose) contains atoms of carbon (C), hydrogen (H), and oxygen (O) in a molecule with interconnected carbons:

$$\begin{array}{c} \text{H H H H H H} \\ |\ \ |\ \ |\ \ |\ \ |\ \ | \\ \text{H—C—C—C—C—C—C}=\text{O} \\ |\ \ |\ \ |\ \ |\ \ | \\ \text{OH OH OH OH OH} \end{array}$$

11. (2) Glacial ice flows slowly as the snow and ice at higher elevation press downward. Glaciers are not frozen rivers but are solid year-round. Today's glaciers are residual from the Ice Age and may be shrinking. The earth becomes warmer at depth.

12. (2) Halley's Comet returns near the sun and earth every 76 years. Its periodic reappearance demonstrates that it has a closed, regular orbit. It is a member of the solar system, orbiting the sun in a very long, narrow ellipse.

13. (4) As long as both ice and water are present in thermal equilibrium, they must be at the melting/freezing point, exactly 32°F. The water temperature could not be lower, nor the ice temperature higher. An armchair experiment with a thermometer in a glass of ice and water will convince you that the temperature remains constant until all the ice has melted.

14. (1) Meiosis is the special mode of cell division that produces haploid gametes, or egg and sperm cells. Mitosis is simple cell-splitting. Mutation is the infrequent modification of a gene. Osmosis is the passage of a solvent through a membrane to dilute a more concentrated solution. Natural selection is the mechanism by which new species arise.

15. (2) MgF_2 salt in solution dissociates to positive and negative ions of magnesium and fluorine:

$$\underset{salt}{MgF_2} \rightarrow \underset{cation}{Mg^{+2}} + \underset{anion}{2F^{-1}}$$

Ions are electrically charged atoms, with the number of electrons not equal to the number of protons. Metallic elements like magnesium tend to lose electrons and become positive cations, while nonmetals like fluorine tend to gain electrons and become negative anions.

16. (1) The ocean floor is littered with limestone, lava fragments, manganese nodules, and oozes rich in diatoms. The limestone was precipitated from seawater, commonly as shells. The lava comes from submarine volcanism. Manganese and other metallic elements dissolved in seawater precipitate to form potato-size nodules which may be an economic ore in the near future. Diatoms are unicellular floating plants with a siliceous, insoluble case. However, an atoll is an organic island in a tropical sea and could not be found in a dredged sample.

17. (1) All organisms manufacture proteins from amino acids. Plants make all their amino acids from still simpler nutrients, but animals always depend on plants to supply some amino acids they cannot make themselves.

18. (2) The force of buoyancy diminishes the apparent weight by the weight of water displaced by the glass object. The relation is known as Archimedes' principle from the incident when that ancient Greek mathematician is said to have jumped out of his bath shouting "Eureka!" after he realized how to measure the purity of a crown. You know from swimming that submerged objects (including people) seem to weigh less.

19. (4) Certainly a hand and a flipper are not used for similar tasks. The possibility that man evolved from fishes is not suggested by the man/whale similarity, for whales are not fishes, but mammals. The proper inference is that man and whale have a common ancestor, the bone structure being inherited from some earlier mammal.

20. (5) The compound DNA is the molecule which records genetic messages. It is found only in the chromosomes of cell nuclei and is duplicated during the cell-splitting of reproduction. Thus genetic information is passed from generation to generation. That information provides detailed control over the development and activity of cells and, hence, entire organisms.

Part B: Science Reading Comprehension

21. (3) Tuberculosis is described as a bacterial infection of the respiratory system, and a tubercular infection tends to persist (is chronic). The disease, however, is not inherited.

22. (3) Death is the direct consequence of lung damage, which is caused by infective bacilli (bacteria). The bacilli produce characteristic nodules, called tubercles, in the lungs, but if the tubercles remain intact, the disease is not serious.

23. (2) *Mycobacterium tuberculosis* is the disease organism. In binomial nomenclature of organisms, the genus is the first, capitalized word and the species is the second word.

24. (4) The passage says nothing about vitamins as prevention or treatment for tuberculosis.

25. (2) Cinders are fragments of volcanic rock spewed out of a volcano during an explosive eruption. They are larger than ash particles but smaller than blocks.

26. (4) Steam and carbon dioxide are gaseous, and the passage states that an abundance of gas causes violent eruptions. Such a volcano would tend to form a cinder cone from the emission of pyroclastic material rather than lava.

27. (2) The fiery cloud (or *nuée ardente*) is an ash flow lubricated with much gas, which allows it to roll downslope at very high speeds. Such a flow buried the city of St. Pierre, on the island of Martinique, in 1902, killing about 28,000 persons.

28. (5) Vesuvius alternates between quiet and explosive activity, according to the passage, and therefore its cone would be built of alternating layers of lava and pyroclastic debris: a stratovolcano.

29. (2) Sound waves are a disturbance traveling through a substance, the elastic medium.

30. (3) The buzzer produces a vibrating disturbance, or sound wave, which travels through the air (if any) and glass of the jar to the surrounding air, where it contacts the observer's ear. The bell is the sound source in the second experiment.

31. (5) As the pumping proceeds, the air pressure inside the jar should fall gradually, so the transfer of sound waves would become slowly poorer.

32. (5) The purpose of the second experiment was to display transmission of sound by a liquid, the water.

33. (4) The cephalothorax comprises the head and thorax of the crayfish, so it is in front of the abdomen.

34. (4) The terminal segment of the abdomen is the telson, which with the uropods constitutes a fanlike tail.

35. (4) The first pair of legs from the front of the crayfish are called chelipeds; they are large pincers.

36. (4) Let's count the *pairs* of appendages from the front: 2 antennae, 1 mandibles, 2 maxillae, 3 maxillipeds, 5 legs, and 6 swimmerets. The 19 pairs of appendages are equal to 38 appendages.

37. (2) Table salt is sodium chloride and is rich in the halogen chlorine. Sodium is a metal, not a halogen.

38. (5) The temperature range in the liquid state is obtained by subtracting the melting point from the boiling point. The greatest range is iodine, 71°. Although bromine is a liquid under normal conditions, its liquid range is only 66°. You should check this last number to see how it was calculated.

39. (5) Normal temperature within a room is about 20°C. The first three halogens melt at a lower temperature. Iodine, however, occurs as crystals at room temperature.

40. (5) Before bromine can be produced by the third reaction, chlorine must be available for the replacing agent. Chlorine is generated commercially by the second reaction, the electrolysis of sodium chloride.

41. (1) Evaporation would cause a lake to diminish or disappear. The other processes are cited as leading to lake basins.

42. (1) Weathering and erosion of the surrounding landscape produces salts dissolved in water. Such salts will become concentrated in lakes with no outlets, which become very saline through time. Some, but not *most,* of the dissolved matter could come from processes (3) and (5). The dissolved material cannot *come from* evaporation (2), which simply concentrates the dissolved salts.

43. (3) Some lakes have large amounts of dissolved borax and other boron salts. Evaporation of such a lake to a playa may leave a mineral deposit.

44. (1) Answers (4) and (5) are not as general as (1): most lakes vanish by filling up with sediment. Salt is only one type of sediment. Lakes in humid regions do not disappear by evaporation or salt precipitation, but by filling up with sand and mud.

45. (4) The passage explains that estrogens cause the endometrium, the lining of the uterus, to grow in thickness.

46. (3) Progesterone secretion begins at ovulation. This is most easily seen on the graph.

47. (1) Estrogens are secreted at an early stage by follicles and then later by the corpus luteum. The two sources explain the two peaks in the graph for the abundance of estrogens.

48. (4) Progesterone is secreted for 16 or 17 days, according to the graph.

49. (3) Population is grainy; that is, the number of people must be integral and not a fraction. The town of Black Hat, Wyoming, must have a population of 202 or 203 persons, but not an intermediate number. The other four measures are not necessarily even, but can be any value.

50. (3) The term classical is used in high praise of the usefulness of Maxwell's theory of electromagnetic phenomena. A classical theory is not necessarily outdated, but it must have endured.

51. (3) Quantum physics states that some quantities are grainy, rather than continuous. The physics of mass, motion, and electricity were established long before Max Planck arrived at the quantum of energy in 1900.

52. (1) The passage states that the quantization of energy can be detected only on the atomic scale, as with the excitement of electrons. That energy can be measured is a principle common to both classical (continuous) and quantum physics. "Particles of energy" would be an awkward oversimplification of quantum theory.

53. (2) Iron pyrite is FeS_2, with one iron and two sulfur atoms per formula unit. So pyrite has 67% sulfur atoms. Chalcopyrite and galena have 50% sulfur, while gypsum has only 1/12, or 8%, sulfur. Note that gypsum, $CaSO_4 \cdot 2H_2O$, has 1 Ca, 1 S, 6 O, and 4 H atoms.

54. (4) Because sulfur melts at only 113°C, it can be melted by superheated water or steam. Thus it can be "mined" by drilling, somewhat in the manner of oil and gas extraction.

55. (5) The deposits in Texas and Louisiana are free (uncombined) sulfur, and thus can be melted. Sulfide or sulfate deposits are mined as solid rock rubble from underground or open-pit mines.

56. (5) The fourth pipe must be the innermost. The air pumped down that pipe serves to force the molten sulfur up the third pipe to the surface. The sulfur is melted by the hot water in the first two pipes.

57. (5) From the passage one can infer that the sporangium is a structure containing spores, and therefore the spores form within the sporangium.

58. (4) The lip cells have thin walls. As the annulus cells shrink, a tear begins to form at the weaker lip cells so the sporangium is bent backward in a manner suitable for hurling the spores later.

59. (2) Shrinkage of the annulus in dry weather bends the "slingshot," the top of the sporangium. The final conversion of water to vapor *releases* the tension and lets the sporangium hurl the spores.

60. (5) Dry weather leads to evaporation of water from the annulus cells, and the consequent shrinkage of the annulus is the engine driving this wonderful mechanism.

FINAL PREPARATION: "The Final Touches"

1. Make sure that you are familiar with the testing center location and nearby parking facilities.
2. The last week of preparation should be spent on a general review of key concepts, test-taking strategies, and techniques.
3. Don't cram the night before the exam. It is a waste of time!
4. Eat a nourishing breakfast before the exam.
5. Arrive in plenty of time at the testing center.
6. Remember to bring the proper materials: identification, admission ticket, three or four sharpened Number 2 pencils, an eraser, and a watch.
7. Start off crisply, working the questions you know first, and then coming back and trying to answer the others.
8. Try to eliminate one or more choices before you guess, but make sure you fill in all of the answers. There is no penalty for guessing.
9. Mark in reading passages, underline key words, write out important information, and make notations on diagrams. Take advantage of being permitted to write in the test booklet.
10. Make sure that you are answering "what is being asked" and that your answer is reasonable.
11. Cross out incorrect choices immediately; this will keep you from reconsidering a choice that you have already eliminated.
12. Using the SYSTEMATIC OVERALL APPROACH is the key to getting the questions right that you should get right—resulting in a good score on the GED exam.

APPENDIX

POLICIES OF STATE DEPARTMENTS OF EDUCATION AND OFFICIAL GED CENTERS

State departments of education issue high school equivalency diplomas, or certificates, on the basis of one or both of the following: (1) the GED score on each of the five tests and (2) the average GED score.

Following is a list of minimum GED score requirements for the various states, provinces, and territories. The information in this table is *correct as of March 1, 1979*. For current information, please contact your local GED Center.

NOTE: If you are using the Practice Tests in this series to predict probable success on the actual GED Test Battery, it is important that you take into account the standard error of estimate, about 8 points on each exam.

The table is courtesy of the GED Testing Service of the American Council on Education.

GED SCORE REQUIREMENTS

State	Minimum Score on Each Test		Minimum Average on All Five Tests
Alabama	35	or	45
Alaska	35	or	45
Arizona	35	and	45
Arkansas	35	and	45
California	35	and	45
Colorado	35	and	45
Connecticut	35	and	45
Delaware	40	and	45
District of Columbia	35	and	45
Florida	40	and	45
Georgia	35	and	45
Hawaii	35	and	45
Idaho	35	and	45
Illinois	35	and	45
Indiana	35	and	45
Iowa	35	and	45

GED SCORE REQUIREMENTS

	Minimum Score on Each Test		Minimum Average on All Five Tests
Kansas	35	and	45
Kentucky	35	and	45
Louisiana	35	or	45
Maine	35	and	45
Maryland	40	and	45
Massachusetts	35	and	45
Michigan	35	and	45
Minnesota	35	and	45
Mississippi	40	or	45
Missouri	35	and	45
Montana	35	or	45
Nebraska	40	or	45
Nevada	35	and	45
New Hampshire	35	and	45
New Jersey	35	and	45 (English) Total 270 (Spanish)
New Mexico	40	or	50
New York	35	and	45
North Carolina	35	and	45 (total of 225)
North Dakota	40	or	50
Ohio	35	and	45
Oklahoma	35	and	45
Oregon	40 on each test		—
Pennsylvania	35	and	45
Rhode Island	35	and	45
South Carolina	—		45 average
South Dakota	35	or	45
Tennessee	—		45 average
Texas	40	or	45
Utah	40	and	45
Vermont	35	and	45
Virginia	35	and	45
Washington	35	and	45
West Virginia	35	or	45
Wisconsin	35	and	45
Wyoming	35	and	45

GED SCORE REQUIREMENTS

U.S. Territories	Minimum Score on Each Test		Minimum Average on All Five Tests
American Samoa	35 on each test		—
Canal Zone	40	and	45
Guam	35	and	45
Kwajalein Island	35	and	45
Puerto Rico	An average standard score of 50 on all five tests with no score below 36 OR minimum standard score on each of five tests as follows: Test 1, 36; Test 2, 42; Test 3, 44; Test 4, 38; Test 5, 46.		
Trust Territory of the Pacific Islands	35 on each test		—
Virgin Islands	35	and	45
Provinces of Canada			
British Columbia	40	and	45
Manitoba	35	and	45
New Brunswick	35	and	45
Newfoundland	35	and	43
Northwest Territories	40	and	45
Nova Scotia	45 on each test		—
Prince Edward Island	35	and	45
Saskatchewan	40	and	45
Yukon Territory	35	and	45